58

퍼블릭 스페이스 in 도쿄

새로운 도시 공간을 위한
코퍼레이티브 디자인

도시 만들기·퍼블릭 디자인 센터
퍼블릭 스페이스 연구회 편저 | 정병균, 김미화 역

58
퍼블릭 스페이스
in 도쿄

새로운 도시 공간을 위한
코퍼레이티브 디자인

【차례】

I — MOBILITY HUB
모빌리티 허브가 되는
퍼블릭 스페이스

II — SEAMLESS LINKAGE
원활한 네트워크를 형성하는
퍼블릭 스페이스

III — URBAN GREEN CORE
주변의 지역과 연계된
퍼블릭 스페이스의 거점

58
Public Spaces in Tokyo

Cooperative Design for
New Urban Infrastructures

【CONTENTS】

I MOBILITY HUB
Public spaces with access
to transportation

II SEAMLESS LINKAGE
A series of spaces that
form a continuous network

시작하며

2020년 전 세계적으로 맹위를 떨친 COVID-19로 인해 의외로 도시에 있어서 퍼블릭 스페이스의 중요성이 재확인되었다. 비상사태 선언으로 자숙이 요청되는 상황 속에서, 혼자 공원의 나무 그늘에서 쉬거나 강변을 산책하거나 조깅을 즐기는 사람이 많이 생겨났다. 또 점포 내 영업이 곤란하게 된 음식점을 위해 도로 공간을 객석으로 사용하는 모습도 보인다. 세계적으로 긴장감이 팽배해진 가운데, 집 근처의 공원에 나가 일광욕을 하고, 바람을 맞으며 산책을 하며, 활력 있는 옥외 활동을 즐기는 것이 몸과 마음의 건강을 유지하는 데 중요한 역할을 한다는 것은 의심할 여지가 없을 것이다.

2022년 2월 현재 COVID-19로 인한 팬데믹의 종식을 예상하기 어려운 상황이며 미래 사회 또한 어떻게 변모할지 예측하기 힘든 상황이 이어지고 있지만, 그럼에도 퍼블릭 스페이스 만들기가 도시에 있어서 하나의 중요한 테마로 자리매김해야 한다는 사실은 의심할 여지가 없다. 본서는 도쿄에 존재하는 퍼블릭 스페이스에 주목했다. 이 고밀화된 도쿄에 양질의 퍼블릭 스페이스가 생겨나는 과정을 꼼꼼하게 살펴보는 것은 미래의 도심환경을 만드는 데 힌트가 되리라 생각한다. 이를 위해 도쿄의 무수한 퍼블릭 스페이스 중에서도 비교적 새롭고 도심에 가까운 사례를 중점으로 소개한다. 대부분은 1990년 버블경제 붕괴 후의 도시재생 프로젝트

에 의하여 생겨난 것들로, 도시에서 생활하는 사람들의 일상생활의 퀄리티를 높이기 위해 공공과 민간이 협력하여 만든 것들이다. 하지만 이 프로세스에는 대립되는 많은 관계자가 생겨났으며 서로 다른 입장을 넘어서 다양한 조율의 과정을 거치면서 비로소 활력 있는 장소로 탄생하게 되었다. 이 조율은 구상단계에서 시작되어 준공과 운영단계까지 긴 시간을 필요로 한다. 여러 관계자의 아이디어가 집적되어 비로소 코퍼레이티브 디자인(Cooperative Design)이 가능하게 되는 것이다.

이렇게 해서 만들어진 퍼블릭 스페이스는 도로, 공원, 상하수도, 에너지 설비 등 공공사업으로써 정비되어 왔던 도시 기반시설의 역할을 보완하는 것뿐만 아니라 성숙한 도시로써 더욱 매력을 높이고 사람들이 생활을 즐기는 활력 넘치는 인간중심의 풍경을 만들어가기 위해 없어서는 안될 인프라스트럭처 그 자체가 되었다. 이 때문에 지금까지의 도시 인프라의 정의를 재해석해 이 퍼블릭 스페이스를 뉴 어반 인프라스트럭처(New Urban Infrastructure)라고 명명하기로 한다.

본서에서는 도쿄에 있는 퍼블릭 스페이스의 구조와 그 형성의 과정을 세밀히 들여다봄으로써 새로운 시대의 도시 만들기의 힌트가 되도록 정리하려 한다.

Introduction

COVID-19 has raged across the world in 2020. This, unexpectedly, has reaffirmed the importance of public spaces in our cities. Although we've been asked to refrain from a growing list of activities during the states of emergency, we still see people out jogging and strolling along rivers and relaxing on picnic sheets amid the greenery of parks. We've also seen proprietors of restaurants and bars, unable to serve customers indoors, using street space for customer seating. With the entire world under strain, there's no doubt that the time we spend on ordinary, human activities outdoors, in nearby public spaces—enjoying the refreshing breeze while basking in the sunlight or walking in the shade of trees—confers important, salubrious effects on our bodies and minds.

At present, in February of 2022, the end of the pandemic is nowhere in sight. While it's hard to imagine all ways society will be affected, the creation of excellent public spaces will surely be recognized as key in future city planning efforts. With this in mind, we decided to devote this issue of the magazinechapter to attractive public spaces that already exist in Tokyo. How were these excellent public spaces realized in this high-density metropolis? The answer may provide hints for future city planning projects, in Japan and other parts of the world. Tokyo has many public spaces, so we will present ones that are relatively new and near the center of the city. Many of these spaces were created through urban renewal projects after 1990, when Japan's bubble economy burst.

These public spaces came into being through collaborations between public and private entities that sought to improve the daily lives of city residents. The collaborations usually involved multiple stakeholders possessing different points of view. Only by transcending their differences, by working through the recriminations of different stakeholders, were lively spaces achieved. In many cases, recriminations continued for exceedingly long periods of time; in some cases, they began in the planning phase of projects and continued beyond the completion of construction, into the operating phase. Since the completed spaces reflect input from multiple stakeholders, they may be considered products of cooperative design.

The public spaces complement other components of urban infrastructure that are provided as public works: roads, parks, waterways, electrical and sewage systems, and so on. They enhance the appeal of Tokyo as a mature city by bringing joy to people's lives and providing a backdrop for lively human drama. For these reasons, public spaces may be considered essential infrastructure. We have therefore redefined urban infrastructure to include public spaces and shall refer to the spaces as "new urban infrastructure."

In this issue of the magazinechapter, we shall introduce examples of these spaces in Tokyo along with analysis of how they were created and how they function. We hope this will provide useful hints for city planning in the post-COVID-19 world.

New Urban Infrastructure의
다섯 가지의 타입

도쿄는 대중교통을 중심으로 건설된 도시이다. 도시 재생을 통해 등장한 많은 퍼블릭 스페이스가 기차역 내부 또는 주변에 있다. 이러한 공간과 역을 연결하면 역에 대한 접근성이 향상되고, 역 주변에서 활발한 활동이 생겨난다. 때로는 한정적으로 사용할 수밖에 없는 공간의 활용성을 높이기 위해 3차원 공간으로 이용하는 사례도 있다. 또한 공간의 활력을 높이기 위해 공간의 운영 단계에서 많은 아이디어를 적용했다. 공간과 그 안에서 일어나는 활동을 통틀어 'New Urban Infrastructure(새로운 도시 기반시설)'이라고 말한다. 이 책에서는 공간의 특성에 따라 5가지 유형으로 공간의 활용 사례를 분류했다.

5 types of
New Urban Infrastructure

Tokyo is a city built around public transportation. Many public spaces that have emerged through urban renewal are in or around train stations. Creating linkages between such spaces and the stations improves access to the stations and also generates lively activities around them. Space in these places is sometimes used dynamically in three-dimensions, to make the most of what little is available. Many ingenious strategies have been employed in the operating phase of the spaces, to increase their vitality. The spaces and the activities that take place in them may be collectively understood as "new urban infrastructure." In this issue we have classified examples of the spaces into five types, according to their characteristics.

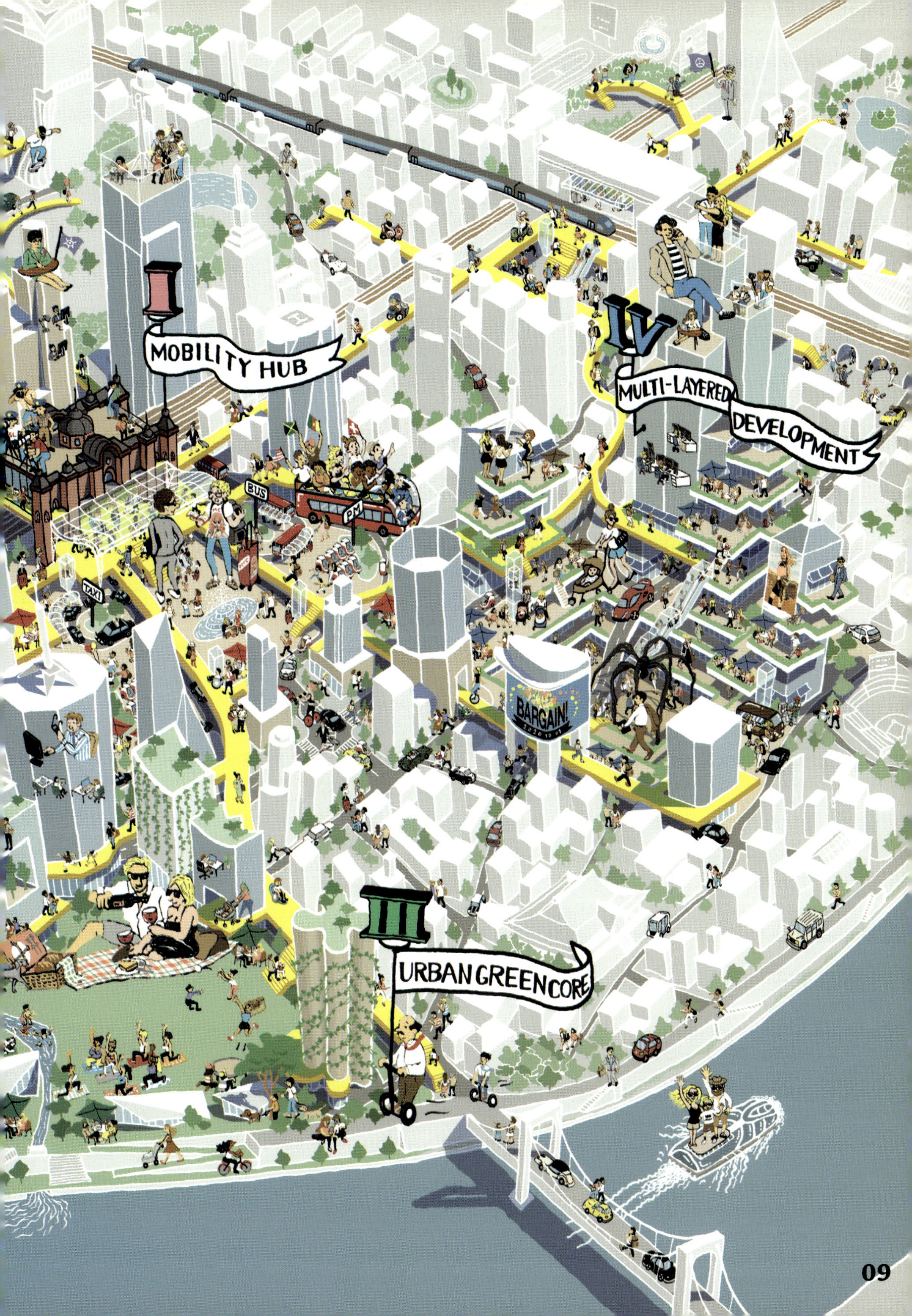

MOBILITY HUB
MULTI-LAYERED DEVELOPMENT
URBAN GREEN CORE
I
IV
III
BUS
PM
TAXI
BARGAIN!
09

뉴 어반 인프라스트럭처의
5가지 타입

MOBILITY HUB

**Public spaces with access
to transportation**

모빌리티 허브가 되는
퍼블릭 스페이스

역주변의 재편에 의한 보행자 공간의 확충

과거에는 역 앞 공간을 주로 버스와 택시 터미널로 사용해 기차와 자동차 사이를 환승하는 사람들을 수용했다. 하지만 최근 몇 년 동안 역 주변 공간이 보행자 중심의 광장으로 재편되는 변화의 움직임이 생겨나고 있다.

민간 개발 및 기반 시설 개선을 포함한 역과 역 주변에 대한 전면적으로 재편되는 프로젝트에 창의적인 솔루션이 사용되었다. 예를 들어, 교통 터미널, 역 앞 중앙광장 및 기타 역 기능의 재편을 민간 개발에 맡겨, 이용자가 다양한 교통수단으로 안전하게 이동할 수 있는 보행자 중심의 퍼블릭 스페이스를 만들 수 있었다. 이러한 변화가 역을 도시로 확장시키는 계기가 되고 있다.

Expanding pedestrian spaces through station reorganization

In the past, spaces in front of stations were primarily devoted to bus and taxi terminals, to accommodate people transferring between trains and automobiles. This, however, has been changing. In recent years, these spaces have been redeveloped as pedestrian-oriented plazas.

Creative solutions have been employed in station-front reorganization projects that involve private development and infrastructural improvements. For example, transportation terminals, concourses, and other station functions have been shifted onto privately developed land. This has made it possible to create pedestrian-centric public spaces that allow people to safely transfer between different modes of transportation. It has also helped the stations to extend out into the city.

II SEAMLESS LINKAGE

**A series of spaces that
form a continuous network**

원활한 네트워크를 형성하는
퍼블릭 스페이스

쾌적한 이동을 뒷받침하는 네트워크

철도, 주요 도로, 하천 등 편안한 이동을 지원하는 선형의 네트워크는 도시의 분할 요소를 뛰어넘어 지속적인 녹지와 활력을 제공해, 사람들이 도시공간을 더 즐겁고 쾌적하게 이동할 수 있게 돕는다. 언뜻 연속된 공간으로 보이는 선적인 네트워크의 내부를 들여다보면 실제로는 서로 다른 소유자의 토지로 구성되어 있다. 따라서 이 공간들을 원활하게 연결하려면, 민관(民官)이 함께 같은 미래 비전을 공유하면서 보행자 공간을 연결해야 한다. 이를 위해서는 명확한 가이드라인과 마스터플랜을 수립해야 한다.

Networks that support comfortable movement

Linear networks extend over railroads, major roads, rivers, and other dividing elements in cities to provide continuous stretches of greenery and vitality. They make moving through cities more enjoyable. At first glance they appear seamless but they actually extend across land belonging to different owners. Creating seamless linkages requires pedestrian spaces to be joined together across multiple public and private developments based on a single, shared vision of the future. Clear guidelines and masterplans must be established to achieve this goal.

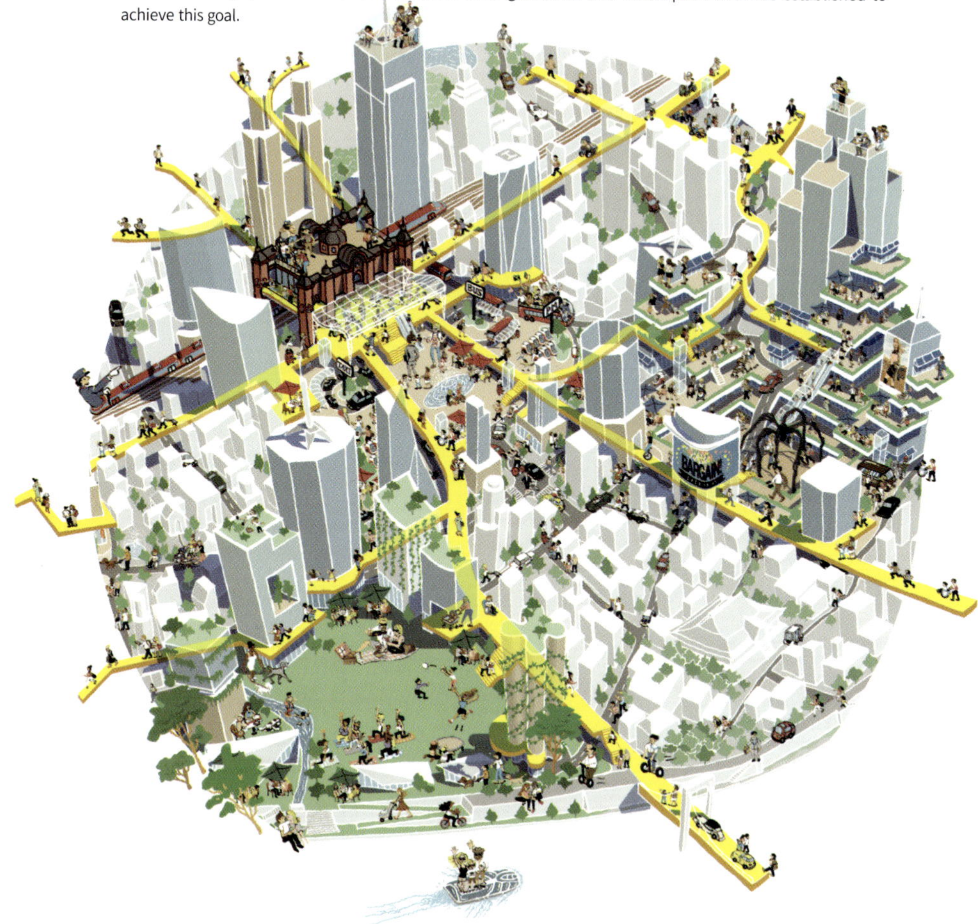

URBAN GREEN CORE

Key public spaces integrated into the surrounding city

주변의 지역과 연계된 퍼블릭 스페이스의 거점

공원이나 하천과 민간개발과의 일체정비

공원, 하천과 민간개발 부지를 포함하는 도시녹지코어(Urban green core)는 도시개발로 만들어진 대규모 녹지 공간 및 수변의 개방 공간이다. 대부분 공공과 민간은 강과 공원 주변에 있는 열린 공간과 민간에 의해 개발된 토지 공간을 결합하기 위해 상호 조정하는 단계를 거친다. 이처럼 민관이 노력을 하나로 합쳐 민간시설의 활력을 퍼블릭 스페이스로 끌어들임으로써 도시녹지코어가 공간의 연속성을 가지게 되며, 도시 보행자 네트워크의 허브로써 역할을 하게 된다.

Improvements involving parks, rivers, and private development

Urban green cores are large green spaces and waterfront open spaces created in conjunction with urban developments. In many instances, government and the private sector coordinate to join open spaces along rivers and parks with spaces on privately developed land. Uniting the efforts of the public and private sectors in this way helps bring the vitality of private facilities out into the public spaces. Urban green cores are designed with consideration of spatial continuity and serve as hubs of the city's pedestrian networks.

IV

MULTI-LAYERED DEVELOPMENT

Conglomerated public spaces crossing
different domains

영역 횡단형
복합 퍼블릭 스페이스

입체적으로 적층한 액티비티

종종 많은 인파가 모이고 활동이 가장 집중되는 도시의 중심은 중층적으로 개발된다. 그 이유는 퍼블릭 스페이스를 입체적으로 배치해 협소한 토지 면적을 최대한 활용할 수 있기 때문이다. 법 개정을 통해 도로, 하천 위와 아래, 건물 위 등의 공간은 관리하는 주체의 경계를 초월해 입체적인 퍼블릭 스페이스로 재탄생된다. 이러한 공간은 도시를 볼 수 있는 새로운 장소를 제공하고, 사람들이 새롭고 독특한 방식으로 시간을 보낼 수 있게 한다. 이러한 입체적인 공간에서 보내는 사람들의 다양한 활동 모습은 도시의 새로운 아이콘이 되기도 한다.

Activity layered in three dimensions

At the center of the city, where large crowds gather and activity is most concentrated, multi-layered developments are often created. Public space is arranged three-dimensionally, to make the most of what little land area is available. Legal reforms have enabled the spaces to unfold three-dimensionally above and below roads and rivers, above buildings, and so on across properties managed by different authorities. These spaces provide new places for viewing the city and allow people to spend time in new and unique ways. Scenes of human activities layered in three dimensions also serve the city as new icons.

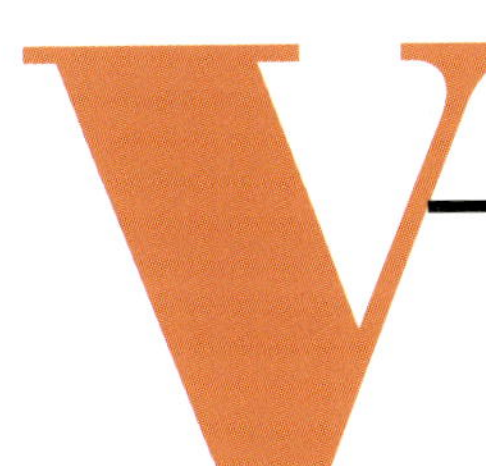

V AREA MANAGEMENT

Public spaces cultivated and operated
by local communities

지역에서 육성하고 운영하는
퍼블릭 스페이스

하이 퀄리티의 운영을 통해 매력을 더해가는 퍼블릭 스페이스

활기찬 활동으로 퍼블릭 스페이스를 활성화하는 데 필요한 것은 공간 디자인이 전부가 아니다. 높은 수준의 관리와 운영시스템도 빼놓을 수 없다. 양질의 유지·관리를 제공하고, 이벤트를 개최하고, 지역 사회와 협력하는 활동을 개최하는 것은 사람들의 생활의 질과 지역 사회와 유대감을 높이는 방법이다. 이러한 활성화는 민간개발 사업자가 단독으로 추진하는 것보다 정부, 지역 주민, 기업 및 사용자의 협력이 필요하다.

Public spaces whose charm is enhanced through high-quality operation

Good spatial design isn't all that is required to animate public spaces with lively activities. Good management and operation are also indispensable. Providing high-quality maintenance and management, staging events, and hosting activities in cooperation with the local community are ways of creating ties to the community that can improve people's daily lives. Such activities are not carried out solely by business operators from private developments; they also involve the cooperation of government, local residents, companies, and the end users.

다수의 주체가 연계되어 같이 만드는 협력형 퍼블릭 스페이스

Collaborative Public Space
Created by
Multiple Entities Working
Together

기시이 타카유키 × 다카미 키미오 × 데구치 아츠시 × 나카이 유

Takayuki Kishii × Kimio Takami × Atsushi Deguchi × Yu Nakai

퍼블릭 스페이스의 변천

— 일본의 퍼블릭 스페이스는 어떤 특징이 있으며, 서양의 퍼블릭 스페이스와는 어떤 차이가 있습니까?

다카미: 일본에 있어서 광장을 말할 때 많은 사람들이 신사(神社)의 경내 등을 떠올리듯, 광장은 일본 문화의 독자적인 것이 아닌 서구의 나라에서 발생된 것이라 할 수 있습니다. 전후 부흥기인 1956년에 만들어진 도쿄 신주쿠 카부키초의 코마 극장 앞 광장은 서구에서 유학한 이시카와 히데아키에 의해 설계된 것으로, 당시 일본에서는 유일하다고 할 수 있는 서양적인 광장이었습니다. 일본에서 서양적인 광장 만들기의 시작은 1964년 해외여행 자유화 시점을 들 수 있습니다. 당시 수많은 젊은이들이 서구 각국을 방문하여 일본에는 없는 광장의 존재를 알게 되었고, 그곳에서 즐겁게 여가를 보내는 모습을 보고 영향을 받은 결과로 일본에 광장 문화가 형성되기 시작되었다고 보입니다. 한편, 본서에서 소개된 58개소의 사례와 같이, 최근에 만들어진 퍼블릭 스페이스의 대다수는 지금까지 광장이 가진 기능에서 멈추지 않고, 우리 세대가 유년시절 백화점의 옥상이나 교외의 유원지 등에서 체험한 것과 같이 여러 사람들과 함께 즐기는 것이 가능한 장소로서의 기능을 가지고 있습니다.

나카이: 서구의 많은 나라들은 18세기 중기 이후 산업혁명과 함께 도시환경의 악화가 진행되었습니다. 노동자들은 위생환경이 열악한 장소에서 고밀도의 생활이 불가피하였고, 그 상황에서 리프레시가 가능한 장소로서 퍼블릭 스페이스가 중요한 역할을 하였다고 할 수 있습니다. 그러나 동시대인 에도시대의 일본은 신사의 경내 등 자연이 풍부한 곳을 거점으로 서민들이 생활하였기 때문에 서구형의 광장 없이도 윤택한 퍼블릭 스페이스의 체계를 가지고 있었습니다. 일본의 도시는 풍부한 자연을 모체로 형성되어 있어 근본적으로 서구 도시들과 공간의 골격이나 구조가 달랐기 때문에 서구의 광장 문화가 뿌리내리기 어려웠다고 할 수 있습니다.

다카미: 벚꽃놀이와 같이 옥외활동을 좋아하는 일본인은 땅바닥에 앉는 습성이 있는 반면 입식 문화의 서양인은 근본적으로 체력과 골격이 다르다고 생각합니다. 최근 일본에 있어서 퍼블릭 스페이스

의 사례를 보면, 옥외에 정비된 경우뿐만 아니라 상업시설과 같이 합쳐진 경우 일부가 옥내화된 경우가 많이 있습니다. 예를 들면 일본 미드타운의 잔디 광장에서 사람들이 잔디밭에 누워 뒹굴며 노는 모습이 많이 보이는 것은 일본다운 특징을 잘 살린 퍼블릭 스페이스의 성공 사례라고 생각합니다. 이처럼 최근에는 서구의 나라들을 모방하는 광장보다 일본인의 생활습관과 문화에 맞춘 퍼블릭 스페이스가 만들어지게 되었습니다.

데구치: 도시의 구조가 다르다는 측면은 일본의 퍼블릭 스페이스가 서구 나라와 달리 독자적인 발전을 하고 있는 하나의 원인이라고 생각합니다. 도시의 구조가 다른 것은 도시를 구성하는 블록의 형태나 사이즈의 차이로 나타납니다. 뉴욕은 거의 균일적인 그리드 패턴이기 때문에 건물이나 오픈 스페이스를 새롭게 만들 경우에는 당연히 같은 사이즈의 블록에서 형성되는 것이 많은 것과 달리, 마루노우치는 에도시대의 부케 야시키(무가 저택)에서 계승된 블록 사이즈로 구성되어 있고, 니혼바시는 상업지역으로 번영한 비교적 작은 블록 사이즈로 구성되어 있습니다. 이처럼 도쿄는 지역에 따라서 다양한 형태의 퍼블릭 스페이스를 생성하고 있습니다.

필수 불가결한 인프라로써의 퍼블릭 스페이스

— 도쿄에 있어서 퍼블릭 스페이스는 어떠한 것이며, 어떻게 시대와 함께 변천되었고, 어떠한 프로세스로 어떻게 변화하였습니까?

기시이: 도쿄에 있어서 퍼블릭 스페이스를 말할 때는, 그것들이 만들어진 계기가 된 개발의 역사나 도시형성의 변천을 살펴볼 필요가 있습니다.

1923년 관동대지진 이후, 도쿄에서는 민간철도회사들의 방사선형 철도개발에 발맞춰 철도노선상에 주택을 개발함으로써 도심과 교외의 주택을 연결하는 철도라고 하는 도시구조가 만들어졌습니다. 또한 시부야, 신주쿠, 이케부쿠로 등 민간철도의 터미널역에서는 철도회사가 역빌딩을 건설해서 백화점을 운영하게 되었습니다. 이렇게 철도 네트워크로 뒷받침된 역 중심 번화가로 인해 현재의 발

The Evolution of Japan's Public Spaces

—**How would you characterize Japan's public spaces? How do they differ from those in the West?**

Kimio Takami: When most people think of Japanese *hiroba* [("*open spaces*" or "*plazas*")], the first thing things that come to mind are the precincts of Shinto shrines. This is because plazas are not native to Japan: they originated in Western countries. The well-known plaza in front of the Shinjuku Koma Theater, in Kabukicho, [*Tokyo,*] was designed by Hideaki Ishikawa, who had studied in the West. When the plaza was completed, in 1956, during the period of postwar reconstruction, it was the only Western plaza in Japan. One thing that may have triggered the creation of other Western plazas was the 1964 lifting of a ban on Japanese people traveling overseas for tourism. Thereafter young Japanese people visited Western countries and discovered plazas. They must have been impressed by lively scenes of Europeans enjoying themselves there. When the young people returned to Japan, they imported the idea of plazas along with other things they'd seen and experienced during their travels. These gradually spread across Japan. The 58 places featured in this magazine reveal that public spaces created in Tokyo in the last few decades are more functionally diverse than the ones created earlier. Some of the newer plazas are places where large numbers of people can enjoy themselves together— places that function like the department store rooftops and suburban amusement parks we enjoyed as kids.

Yu Nakai: In Western countries, urban environments began to deteriorate in the mid-eighteenth century as a consequence of industrialization. Laborers began living in close quarters without adequate sanitation, in slum-like residential environments. There was a great need for public spaces where people could relax. Meanwhile, in Edo-period Japan, ordinary Japanese people were free to amuse themselves in the precincts of nearby shrines or places of scenic beauty. Though Japan lacked Western plazas, its urban environments were blessed with other sorts of public spaces. It's worth mentioning that Japan's cities developed out of rich natural environments, which can't necessarily be said of cities in other countries. Given these differences, it's understandable that Western plazas never really caught on in Japan.

Takami: Japanese people are accustomed to sitting on the ground, as they do during *hanami* [("*cherry-blossom viewing*")]. Westerners, on the other hand, seem more inclined to remain standing. Their physiques are different from those of the Japanese. If one looks at recent examples of public spaces in Japan, one will notice that not all of them are outdoor spaces—many are at least partially indoors and were created in skillful combination with commercial facilities. The Garden Lawn at Tokyo Midtown, a place equally suited to playing and relaxing, is an example of a public space that exhibits Japanese characteristics. It proves that Japanese are able to create public spaces that agree with their lifestyles and physiques—that they needn't imitate the plazas of Western countries.

Atsushi Deguchi: The distinctive origins of Japanese cities is one factor that has enabled them to develop in ways that differ from European cities. The origin of any city is manifested in the forms and sizes of its blocks. New York City, for example, is based on a fairly homogenous grid pattern. When new buildings and open spaces are created there, they must fit within the city's similarly sized blocks. As a result, these buildings and open spaces tend to conform to conventional types. Things are different on the eastern side of Central Tokyo. For example, blocks in the Marunouchi area are large, based on spacious samurai residences of the Edo period while blocks in the Nihonbashi area, a renowned merchant district, are much smaller. When the eastern side of Central Tokyo underwent land readjustment work after the Great Kanto earthquake [*of 1923*] and again after the war, different standards were flexibly applied. The heterogenous block sizes in different parts of Tokyo give the city's public spaces distinctive forms and contribute to their diversity.

Public Spaces as Essential Infrastructure

—**How would you characterize Tokyo's public spaces? How have they evolved over time? What changes have occurred in the processes by which they are created?**

Takayuki Kishii: To properly discuss public spaces in Tokyo, we must examine the history of the development projects that generated the public spaces. We must also examine the evolution of the ways in which the city has formed.

After the Great Kanto earthquake of 1923, private railway companies established railway lines radiating out from Tokyo. Land along the lines was developed for housing. This yielded an urban structure where the city center and suburban housing are linked by railways. The private railway companies constructed station buildings at the

전된 도시의 구조를 만들게 되었습니다. 도쿄에서 역은 도시활동의 결절점으로써 매우 중요한 공간이라 할 수 있습니다. 또한 전후 1964년의 도쿄 올림픽 당시에는 모터리제이션에 대응하기 위한 간선도로의 정비가 진행되었습니다. 특히 시부야 주변에서는 메인 스타디움이 있는 국립경기장과 세타가야구 코마자와의 제2경기장을 연결하는 국도 246호가 정비되어 지금의 요요기공원에 선수촌이, 그리고 그 선수촌을 연결하기 위한 수도고속도로와 환상 7호선의 정비가 진행되었습니다. 1970년대에는 니시신주쿠의 요도바시 정수장 부지에 수많은 초고층빌딩이 건설되면서 이케부쿠로에서는 도쿄구치소 전 부지에 썬샤인 60이 계획되었습니다.

그 후 버블경기 즈음에는 경관 및 디자인이 사람을 끌어모은다는 인식이 일반화되어 각지에「특색 있는 경관만들기」가 활발하게 진행되었는데, 지금 돌아보면 믿기지 않는 과도한 개발과 퍼블릭 스페이스가 만들어졌습니다. 이러한 상황에 위기감을 느낀 각 분야의 전문가는 도시경관을 재정비하기 위한 체제를 만들기 위해 연계하여 1989년 도시환경 디자인에 관한 공익재단법인 도시만들기 퍼블릭 디자인 센터(udc)를 설립하였으며, 1991년에는 도시환경디자인에 관한 전문가가 모여 도시환경디자인협회(JUDI)를 조직하였습니다. 현재, 1960~1970년대에는 시부야, 신주쿠, 이케부쿠로에 개발된 건물의 갱신 시기를 맞이해 많은 재개발이 상당히 활발하게 진행되고 있습니다. 도쿄의 경우 도시재생긴급정비지역으로 지정된 지역은 대다수 철도역 주변이었습니다. 공공교통망에 의해 유지되고 있는 도쿄는 역 주변의 중요성을 인식하여, 본서에 등장하는 퍼블릭 스페이스의 대부분을 역 주변에 만들었습니다.

퍼블릭 스페이스는 도시개발의 변천과 함께 발전해 왔습니다. 퍼블릭 스페이스는 원래「공공의 공간」이라는 의미로써 도로, 하천, 공원 등의 공공시설, 즉 행정이 관리하는 일차적인 인프라를 의미합니다. 단지 도쿄에서는 철도, 그 중에서도 역은「모두의 공간」으로 인식됩니다. 또 일본에서는 개발 프로젝트 중에서 공공공간을 만들면 용적률 완화의 인센티브를 받을 수 있는 제도가 있기 때문에 도심 곳곳에 작은「공개 공지」가 생겨났습니다. 다만, 시대의 변화와 함께 가치도 변해 이러한 다양한 개별의 공간을 연결해 보다 풍부한 퍼블릭 스페이스를 만드는 것을 지향하게 되었으며, 최근에는「활력을 만들어내는 장치」로 발전한 퍼블릭 스페이스도 적극적으로 만들어지게 되었습니다.

데구치: 저의 은사님인 오타니 유키오 선생은 자신의 저서「공지의 사상」(호쿠토출판, 1979년)에서 초고층 빌딩 건설에 의해 도시가 고밀도화된 1970년대에는 퍼블릭 스페이스를 만드는 것이 도시계획의 역할이라고 주장했습니다. 그 책에서는 고밀도화 되어 가는 도시에 있어서 보이드 공간의 필요성을 말하고 있으나 그것은 어디까지나 고밀도화에 대한 억제력을 만들고 채광과 통풍 등 환경적 장치로써의 역할이 크다고 주장하고 있습니다. 그러나 현대에 있어서 퍼블릭 스페이스는 고밀도의 활력을 만들어내는 역할을 담당하는 것으로 인식되는 경향이 있고, 오타니 선생이 약 40년 전에「공지의 사상」에서 지적한 공지의 역할과는 반전된 상황이라 할 수 있겠습니다. 사회적 의미나 담당하는 역할이 변화하고 있기 때문에

당연히 디자인 사고도 변화되어 왔습니다.

기시이: 오픈 스페이스에 요구되는 배경도 변화되고 있습니다. 약 30년 전부터 도쿄 다이마루 지역에서부터 재개발이 시작되어, 그 뒤로 롯본기, 토라노몬, 최근에는 시부야, 신주쿠, 이케부쿠로에서도 재개발이 진행되면서 다양한 형태의 퍼블릭 스페이스가 정비되고 있습니다. 각 지역이 도쿄의 중심부 기능을 담당하는 지역으로 발전하고 있지만, 각각 장소와 개발 시기가 다르기 때문에 도시의 갱신에 관해서 도쿄는 유연하게 대응하는 것이 가능합니다. 어떤 지역에서는 재개발이 진행되는 반면 다른 지역에서는 경제활동이 가능합니다. 이러한 지속적인 갱신의 실현이 가능한 도시구조는 도쿄의 장점이라고 할 수 있을 것입니다. 더욱이 도쿄에서는 도시 만들기를 하나가 아닌 다수의 주체가 협력해 기반을 다지고, 경우에 따라서는 부지를 교환하면서 퍼블릭 스페이스를 만들어가고 있습니다. 즉 여러 공간요소를 묶어 연결을 구축하는 것으로 일체적인 퍼블릭 스페이스를 형성해 지역의 가치를 높이는「협조형 디자인」방식이 공유되고 있습니다.

데구치: 지역의 가치를 높이기 위해 최근에는 대형 부동산 디벨롭퍼나 철도사업자가 자신들이 부동산을 소유하는 특정지역에서 그 특징을 살린 개발을 하게 되었습니다. 비슷한 방식의 모델로써 다이칸야마 힐사이드 테라스(1969~1998년)의 성공의 사례를 들 수 있습니다. 이는 로컬 디벨롭퍼 아사쿠라 부동산과 건축가 마키 후미히코가 시간과 정성을 들여 매력적인 건축과 퍼블릭 스페이스를 만든 선구적인 사례로, 소규모이지만 세련된 건축과 다양한 오픈 스페이스가 동시에 디자인되어 양질의 커뮤니티 공간이 형성되어 있습니다.

나카이: 정부/행정 주도로 진행되는 도시만들기는 일정한 제도나 기준, 관리에 한정되는 경향이 있기 때문에 비슷한 정도의 성과에 머무르게 됩니다. 그러나 최근 도쿄에서의 재개발과 같은 민간 주도 방식은 건축과 일체화되어 다양한 퍼블릭 스페이스를 실현할 가능성을 보입니다. 도쿄라고 하는 도시는 하나의 캐릭터나 이미지로 전체가 통일되어 있다기보다는 개개의 지역과 지역의 특징을 연결해서 만든 패치워크와 같은 집합체라고 할 수 있습니다. 도쿄에 있어서 퍼블릭 스페이스도 틀에 맞춘 경직된 방식이 아닌 변화와 성장의 여지를 남겨둔다면 더욱 도쿄다운 다음 세대의 도시 공간으로 진화되어 갈 것입니다.

다카미: 한편 지정용적률 이상을 필요로 하는 개발이 진행되는 지역은 일본 전역을 보더라도 그리 많지는 않습니다. 즉 도쿄에 있어서 퍼블릭 스페이스의 개발 방식이 지방 등 다른 소도시에서 어느 정도 응용이 될지 불확실한 부분이 많습니다.

나카이: 어떻게 해서 퍼블릭 스페이스를 만드는지의 방법론은 가능한 다양해야 합니다. 특정 구역제도나 종합설계제도 등을 활용하는 경우, 많은 사례에서는 규제완화로 인해 얻은 할증용적에 의해 늘어난 바닥면적으로 이익을 취하는 것으로 투자자금을 회수하는 방법이 채택됩니다. 즉, 도쿄 도심의 많은 퍼블릭 스페이스는 규제완화로 인해 얻은 이익과 물리적인 도시를 개혁하는 것으로 탄생됩니다. 그 수법의 일반성을 묻기보다도 그 이외의 선택지를 고려할

railway terminals, in Shibuya, Shinjuku, and Ikebukuro. These buildings contained department stores, which were managed by the private railway companies and supported the private railway networks. This arrangement fostered Tokyo's present-day structure, where bustling shopping and entertainment districts surround the major stations. In Tokyo, stations are crucial places that serve as the hubs of urban activities. After the war, around the time of the 1964 Tokyo Olympics, a network of highways was created to accommodate increasing traffic. National Route 246 was built near Shibuya as a link between the Yoyogi National Gymnasium and the Komazawa Olympic Park Stadium. Nearby, the Olympic Village was built on a site that later became Yoyogi Park. Finally, the Shuto Expressway and Kannana-dori [("Ring Road No. 7")] were built to link up all of the above. In the 1970s, a multitude of high-rise buildings were erected on the former site of the Yodobashi Water Purification Plant, in Nishi-Shinjuku, and the Sunshine 60 Building was built on the former site of Sugamo Prison, in Ikebukuro. During the era of Japan's bubble economy, architects and city planners reawakened to the idea that designs and landscapes could be used to bring people together. Attempting to realize distinctive landscapes wherever possible, they created developments and public spaces that seem outlandish to us today. Experts from different fields observed this situation and were struck by a sense of impending crisis. In 1989, the Urban Design Center, an organization devoted to the design of urban environments, was established to provide a framework for studying cityscapes. Next, in 1991, Japan Urban Design Institute (JUDI) was established as a professional organization for connecting specialists involved with urban design. Many redevelopment projects are now underway in Shibuya, Shinjuku, and Ikebukuro because the buildings were built in 1960s and '70s and have grown obsolete. Since urban renewal was given "nationally strategic" status, private-sector development has been proceeding vigorously. In the case of Tokyo, the areas designated as "Urban Renaissance Urgent Redevelopment Areas" were, unsurprisingly, the areas around railway stations. Since Tokyo is propped up by its public transportation network, it's only natural that these areas would be the most valuable. Most of the public spaces featured in this magazine are also situated near stations.

Japan's public spaces diversified over time as its urban development projects evolved. First of all, there was the basic infrastructure managed by the government: roads, rivers, parks and other public facilities, referred to as "paburiku supesu"[("public spaces")]. Next, in Tokyo, there were the railways and railway stations, which aren't public but have been regarded as "minna no kukan" [("spaces belonging to everyone")]. Next, Japan put in place systems that ease restrictions on the floor area ratios of development projects when the projects include public open spaces. The systems generated lots of small "kokai sareta kuchi" [("spaces open to the public")] in popular destinations across the country. Our values changed over time and we began linking up these discrete spaces to form richer ones. Finally, today, we're proactively creating more evolved public spaces that are capable of "drawing crowds."

Deguchi: Professor Sachio Otani, my former teacher, wrote a book called Kuchi No Shiso [(literally, "Thoughts on Open Spaces")] (Hokuto Shuppan, 1979) in the late seventies, when urban density was increasing due to skyscraper construction. In the book, he asserts that the creation of open spaces is within the purview of urban planning. He advocates the necessity of void spaces in highly dense cities, not only for checking further densification but, more importantly, for the positive effects they confer in securing access natural light and ventilation. Today, however, people expect contemporary open spaces in cities to draw crowds. One could argue that this is contrary to what Professor Otani wrote Kuchi No Shiso some 40 years ago. As the social significance and roles of open spaces have changed, so too has thinking on how the spaces ought to be designed.

다카미 키미오. Mr. Takayuki Kishii

Kishii: I agree with what you are saying. Another aspect that's changed are the settings in which people expect open spaces. Redevelopment in the Daimaruyu area [(the area around Tokyo Station, comprising Otemachi, Marunouchi, and Yurakucho)] began roughly 30 years ago. It was followed by redevelopment in Roppongi and Toranomon, and, more recently, by the ongoing redevelopment in Shibuya, Shinjuku, and Ikebukuro. These redevelopment projects have generated many new public spaces. Each area has evolved into a place that performs functions critical to the city, yet Tokyo as a whole has responded flexibly to their renewal. This was possible because the areas are discrete and their redevelopment was not undertaken simultaneously. When any one area underwent redevelopment, economic activities in the other areas continued, unimpeded. One of Tokyo's strengths is this urban structure, which enables continuous renewal. As a further matter, no single entity is responsible for creating the city of Tokyo; multiple entities cooperate to improve its basic infrastructure. Sometimes these entities even swap building sites to create public spaces. By bundling together various spatial elements and building connections between them, the entities generate public spaces that are unified. This increases the overall value of the areas and reflects a shared understanding of "cooperative design."

Deguchi: Major real estate developers and railway companies seek to increase the value of areas in which their assets are located by highlighting special qualities of the area. A successful example of this approach is the Hillside Terrace Complex (1969–1998), in Daikanyama. It's a pioneering example of a project in which a local developer, Asakura Fudosan, and a single architect, Professor Fumihiko Maki, carefully collaborated over many years to create an attractive stretch of cityscape. Though the project is small in scale, Maki managed to create high-quality communal spaces by designing refined buildings in combination with a variety of open spaces.

Nakai: When the government engages singlehandedly in community development, it tends to get hung up on rules, standards, and management considerations. Its results tend to look pretty similar. Private developers on the other hand have recently undertaken numerous redevelopment projects in Tokyo and have achieved a much wider range of results, often by creating buildings in combination with public spaces. Tokyo does not have a character or image that

필요가 있을 것입니다. 방법론을 넓혀가면 보다 다양한 공간문화가 자라날 것이라고 생각합니다.

콜라보레이션하는 지금부터의 퍼블릭 스페이스

― 다음 세대의 퍼블릭 스페이스에서는 어떠한 것이 요구되며, 그 계획에 있어서 어떠한 것이 중요하다고 생각합니까?

기시이: 이전 독일의 방송국에서 도쿄의 취재를 위한 안내를 의뢰받은 적이 있었습니다. 그 취재진에 도쿄 미드타운의 잔디 광장을 소개했을 때, 도심에도 이처럼 넓고 풍부한 녹지가 남아있는 것에 놀라는 것을 보았습니다. 그러나 실제로는 단순히 남아있는 것이 아닙니다. 도쿄에서 고도집중화가 진행되면서 도심에 있던 정부기관을 타지역으로 이전시키고 그곳의 토지를 개발하여 잔디 광장을 만든 것입니다. 따라서 공공기관을 의도적으로 분산시킨 결과이지, 우연히 퍼블릭 스페이스가 남겨진 것이 아닙니다. 현재, 도쿄에서 진행되고 있는 재개발의 다수가 이러한 맥락에서 이루어지고 있다고 할 수 있습니다. 우선 새로운 여백(토지)을 만들기 위한 구상이 출발점이 되어 도시재생의 실현이 가능하게 된다는 것을 인식하고, 지금부터 의식적으로 도시의 여백 만들기에 착수할 필요가 있습니다.

데구치: 도쿄 23구 전체에서 개발의 균형을 취할 필요가 있다고 생각합니다. 도쿄에 있어서 개발은 남측 구역에 집중되고 그 외의 지역은 둔화의 경향이 있습니다. 도쿄라고 하는 도시 전체의 캐퍼시티를 확장하기 위해서는 북측에 위치한 우에노나 오지 등 거점으로써 기능하는 철도역이나 버스 네트워크가 집중해 있는 지역에 새로운 퍼블릭 스페이스를 만드는 등으로 해서 북측 지구를 재생하는 것이 지금부터의 과제라고 생각합니다. 그 개발사업의 내용도 남측과는 다를 것이며, 때문에 지금과는 다른 플레이어가 요구될 것입니다.

기시이: 그렇군요. 그리고 그 개발은 조금 전에도 언급했습니다만, 지역 전체로써 연결을 가지고 콜라보레이트로 가치를 높여가는 협조형 퍼블릭 스페이스가 요구되고 있습니다. 따라서 건축 개개가 아닌, 지역 전체를 연계한 지역 매니지먼트가 또한 필요한 이유입니다.

데구치 아츠시. Mr.Atsushi Deguchi

데구치: 장기간에 걸쳐 같은 지역의 개발을 점증적으로 지속해가는 일본의 디벨롭퍼나 철도사업자는 세계적으로 봐도 드물다고 하겠습니다. 그들은 지역의 에셋 홀더이며, 지역 전체의 가치를 높여가는 것이 자신들이 소유한 부동산의 가치를 높이는 것이 되므로 지역의 발전을 주도해가는 것입니다. 그 존재는 크고 지역을 대표하는 기업으로써 리더십을 발휘해 개발 매니지먼트를 주도함으로써 지역 전체의 활성화를 도모합니다. 이러한 현상은 디자인의 저변에 있어 눈에 보이지 않지만 협조형 개발 시스템의 하나라고 하겠습니다.

다카미: 본서에 게재되어 있는 다수의 퍼블릭 스페이스는 공공용지와 민간토지를 포함한 협조형 공공공간이고 개발 주체가 상응의 투자를 해서 정비된 것입니다. 사업자가 그것들을 관리하고 있기 때문에 종래보다도 더 다양한 형태로 활용되는 경우가 많지만, 그 관리체제에 일방적인 의존은 금해야 할 것입니다. 유럽의 광장은 어디든지 사람들의 활력이 넘치는 성공적인 공간으로 보이지만, 이는 몇 백 년에 걸쳐 시행착오를 반복하면서 이용법과 관리법을 모색하고 수정 보완한 결과라고 생각합니다. 수많은 도쿄의 퍼블릭 스페이스를 관리하려는 노력뿐만 아니라 이미 있는 공간으로 퍼블릭 스페이스를 사용하려는 움직임도 필요하다고 생각합니다. 이러한 움직임이 일반화되어 가면 더욱 다양한 전개가 생겨날 것입니다.

나카이: 공개 공지로써 정비된 많은 퍼블릭 스페이스에 금지사항이 적혀 있는 것을 목격합니다. 예를 들면, 애완견과 산책금지, 고성금지, 무단촬영금지, 집회금지 등입니다. 자유로운 산책도 못하는 도시공간이, 자유롭게 사진도 찍지 못하는 공간이 시민을 위한 퍼블릭 스페이스라고 할 수 있는지, 정말로 도시재생이나 도시만들기 일환의 역할을 하는지 의문입니다. 시민들이 참여자로서 애착을 갖지 않는다면 누구도 관심을 가지지 않고 관리되지 않는 데드 스페이스가 되어버릴 것입니다. 사람들의 자유로운 활동과 체류가 도시의 매력과 활력에 있어서 불가결의 영양분입니다. 도쿄의 경우 누구든지 자유롭게 이동할 수 있는 교통 인프라가 수준 높게 정비되어 있어 역을 중심으로 사람들이 체류하고 교류할 기반이 갖춰져 있습니다. 그에 더해 이제부터는 중심가로, 보행로, 골목길을 포함해 역과 역을 연결하는 지면 레벨의 도로 네트워크 등을 생각할 필요가 있습니다. 성숙한 사회 속에서 차후 퍼블릭 스페이스에 요구되는 것은 그와 같은 자유로운 액티비티에서 생겨나는 다양한 가치가 아닐까 생각합니다.

(2020년 3월 10일, 니켄세케이 다케바시 오피스에서, 정리: 본지편집부)

코로나 팬데믹으로 인해 재조명되는 퍼블릭 스페이스의 역할과 가치

― 3월에 열린 좌담회 이후 신형 코로나바이러스 감염증이 확대되고, 그 영향을 받은 생활양식이 변화되고 있습니다. 이러한 상황을 거쳐 앞으로의 퍼블릭 스페이스는 어떻게 변화될 것으로 예상하십니까?

기시이: 우선 With 코로나과 After 코로나로 고려해야 할 항목이 다

is consistent throughout the city; it is a patchwork aggregation of distinctive areas and districts. As public spaces in Tokyo grow less formal and less conventional, as they become increasingly diverse, we draw nearer to a new era in which Tokyo's urban spaces better reflect the character of the city.

Takami: Besides Tokyo, though, only a few areas in Japan have development projects underway that exceed floor area ratios designated in the areas. That is to say, I'm not convinced that discussions of public spaces in Tokyo are applicable to the countryside or other areas [in Japan].

Nakai: Yes, I suppose we ought to have a greater variety of methodologies for creating public spaces [in Japan]. Most projects that apply the Specific Block System*1 or Comprehensive Design System*2 rely on a method where additional floor area is obtained through eased regulations is sold to recover investment capital. In a nutshell, many of the public spaces in Central Tokyo, which have physically altered the city, were created with profits gained through eased regulations. Rather than question the universality of this method, we ought to devise other options. A wider range of methodologies will no doubt foster greater diversity in our spatial culture.

Public Spaces of the Future Will Be Collaborative

—What will be expected of public spaces in the future? What kinds of things will be important in planning these spaces?

Kishii: A while back, a German television station reporting on the Tokyo cityscape asked me to serve as a guide. The reporter who introduced the Garden Lawn at Tokyo Midtown remarked, "I'm surprised that such a large, green space remains here, in the center of the city!" In truth, the space hadn't simply remained there. As Tokyo grew more and more concentrated, governmental agencies in the city center relocated to other areas, which freed up land for redevelopment. The creation of the Garden Lawn was one upshot of this process. So, in fact, the public agencies were dismantled deliberately and the public space wasn't there by chance. Many of the redevelopment projects that are now underway in Tokyo arose in a similar manner. Urban renewal is only achieved when movements create land to work with. In the future, we will have to work consciously to create such movements.

Deguchi: I think it will be necessary to promote more balanced levels of development across all 23 of Tokyo's wards. Past development can be characterized as "high in the south, low in the north." That is to say, most large development projects were undertaken on the south side of the city. To increase Tokyo's overall capacity, the next task will be to redevelop its north side. This can be achieved by creating new public spaces and so on by railway stations-for example, in Ueno or Oji- or in areas where bus networks are most concentrated. Since the details of these development projects would differ from those of projects on the south side, different participants would most likely be involved. One can think of many parcels of land that are ripe for redevelopment.

Kishii: I agree. And as I mentioned earlier, any such developments will benefit from cooperative public spaces that connect with and add value to the local communities based on collaborations. For this to be achieved, the public spaces will have to be maintained through [what is referred to in Japan as] "area management." In other words, they will have to be managed not by single entities, but cooperatively, by different community groups from the areas.

Deguchi: Japanese developers and railway companies tend to remain involved with specific areas over long periods of time. This isn't typical in other parts of the world. It happens in Japan because certain companies have considerable real estate holdings in certain areas. The companies are able to increase the value of these holdings by increasing the value of the broader areas in which they are situated. The companies retain a strong presence in the areas and take leadership in managing the development that goes on there, which has the effect of revitalizing the areas. This is how things work behind the scenes in one mode of cooperative design.

Takami: Most public spaces featured in this issue are cooperative ones composed of both public and private lands. They were created by business enterprises that invested appropriate amounts of capital. Since the private sector handles the management of these spaces, they tend to be used in more innovative ways than those maintained by the government. However, we may not be able to rely on this management structure in future development projects. Although European plazas bustle with people and appear successful, we must recall that their management today is the result of hundreds of years of trial and error. What we see is the fruit of those labors. With public spaces in Tokyo, the companies that physically build the spaces typically go on to manage them. However, I think it would be great if there were a movement to transform existing spaces into public spaces rather than just creating and managing new ones. If such a trend took hold, we could expect to see further evolution of the spaces.

Nakai: Open spaces created by developers based on eased regulations tend to have more rules posted than other types of public spaces. For example: "no dog walking," "no yelling," "no napping," "no unauthorized photographing," "no congregating." Can city spaces designed to prevent trouble at any cost, that forbid people from walking their dogs and taking pictures, truly be described as public spaces for the citizenry? Can such spaces play a meaningful role in community development? When citizens are unable to engage with spaces or develop attachments to them, the spaces become poorly managed "dead spaces" that belong to no one. People's freedom to move around and congregate is essential to the charm and vitality of cities. Tokyo has a world-class transportation infrastructure that enables its people to go where they please. In fact, the backbone of the city is its train stations, which serve as hubs for congregating and mingling. Since Tokyo is already well provided with public spaces adjoining stations, as is clear from the examples in this magazine, it's time we think on a grander level about roads connecting the stations-about main streets, promenades, and even alleyways. What will be required of public spaces in the future of our mature society is the diverse values that develop when citizens are free to move around and congregate as they please.

(March 10, 2020, at Nikken Sekkei Takebashi Office;

edited by Shinkenchiku-sha Editorial Department)

Questioning the Role and Value of Public Spaces in Light of the Coronavirus Pandemic

—Since our last round-table discussion in March, COVID-19 has spread broadly and our lifestyles have changed. How has this changed your thinking about the public spaces of the future?

롭니다. With 코로나는 한정된 기간으로 예상되므로 여기에서는 After 코로나에 대해서 토론하는 것이 좋을 것 같습니다. COVID 19 바이러스 감염증은 비말감염과 접촉감염 등 사람의 신체를 통해서 감염되기 때문에 사람들의 접촉과 활동을 규제하는 대책이 마련되고 있습니다. 재택근무나 온라인 강의 등 생활패턴에 변화가 있습니다만, 현재의 감염자 수를 보더라도 사람과 사람의 접촉을 완전히 차단하는 것은 어렵다는 것을 알 수 있습니다.

저는 리모트 워크에 한계를 느끼고 있습니다. 사람과 직접 만나고 정보를 교환하는 것에 가치가 있다고 생각합니다만, 그럼에도 불구하고 앞으로의 생활양식은 크게 변화할 것 같습니다.

다카미: 지금까지는 사람들이 밀집해 고밀화된 도시활동이 일반적이었습니다만, 지금부터는 여러 경우에서 적절한 밀도를 고려할 필요가 있습니다. 지방의 가치가 높아지고 일극집중의 도쿄 도심부의 밀도가 저감되어 점차 전체적으로 균일화된 도시구조가 될지도 모르겠습니다. 한편 지금까지의 도심 퍼블릭 스페이스는 휴식이나 이벤트의 장소로써 역할을 담당해 왔습니다만, 지금부터는 더욱 더 다양한 목적에 대응하는 것이 요구될 것입니다. 차후로는 리모트 오피스로써 몇 시간 그곳에서 일하는 사람들이 생겨날지도 모릅니다. 또한 교외의 공원 다수는 현재 휴일에 여가나 스포츠를 즐기는 장소로 설계되었지만 지금부터는 일상적인 이용이 요구될지 모릅니다.

데구치: 리모트 워크의 도입에 의한 재택근무의 시간이 늘어났지만 기분 전환으로 주위를 산책하며 공원에 가보면 저와 같이 리모트 워크에 지친 사람들이 휴식을 취하거나 아이들과 함께 시간을 보내는 등 다양한 광경을 볼 수 있습니다. 퍼블릭 스페이스는 본래 같은 공간에 있으면서 서로 다른 사람들이 공존하는 가운데 독서를 한다든지 생각에 잠긴다든지 자유로운 시간을 보내는 것을 보증하는 장소였습니다. 저는 지난번 좌담회에서 "현대에 있어서 오픈 스페이스는 도시의 활기를 만들어내는 역할을 담당하고 있다"라고 말했습니다만, 활기를 만들어내는 장치로써의 퍼블릭 스페이스 역할에만 얽매여 있었다는 생각이 듭니다. 다카미 선생이 말씀하신 것처럼 지금부터는 주택지의 공원 등 사람들 생활에 가까운 퍼블릭 스페이스의 역할이 재정립되어가는 과정이 필요하다고 느낍니다.

나카이: 그렇습니다. 저도 리모트 워크를 경험했습니다만, 통근과 교외거주라고 하는 현대도시의 일하는 방식과 생활하는 방식은 겨우 100~150년 상간에 형성된 것일 뿐임을 새삼스레 깨달았습니다. 일본은 에도 시대 말기에서 다이쇼

다카미 키미오. Mr. Kimio Takami.

시대에 걸쳐 콜레라, 장티푸스, 결핵 등의 감염증 유행을 여러 번 겪었고 이로 인해 많은 사람들이 목숨을 잃었습니다. 예전에는 우물물을 통해 감염이 확대되지만, 코로나 바이러스 감염증은 우리들이 만들어내고 있는 밀폐상태에 의해 감염리스크가 더 높아지고 있습니다. 근대에서 현대에 걸쳐 도시는 공간을 구분하고 밀폐해서 그 속의 기능이나 환경을 정교하게 컨트롤하는 것으로 인해 발전해 왔습니다. 그러나 이제부터는 코로나 바이러스로 인해 사이버공간(가상공간)의 가능성이 확대되고, 그에 따라 인터넷 등이 우리 인간의 행동을 보다 고도로 컨트롤하게 될지도 모르겠습니다. 빛과 신선한 공기를 느끼며 사람들과 대화를 하는 가치는 앞으로도 변하지 않을 것이고 퍼블릭 스페이스의 존재가치는 바로 거기에 있는 것이라고 생각합니다. 보다 자유로운 실공간을 획득할 수 있을지가 시험되고 있는 상황이라 할 수 있습니다. 이 기회에 우리들이 당연하게 생각하고 있는 일하는 방식, 생활방식을 새롭게 재해석하는 것으로 보다 자유로운 도시와 퍼블릭 스페이스의 원리가 보일지도 모릅니다.

기시이: 철도가 탄생되면서 환경이 악화된 도심을 떠나 교외로 주택지를 만드는 「직주분리(職住分離)」가 추진되었습니다. 그러나 고도정보화 네트워크의 진전으로 인해 직주분리와는 다른 세계가 출현했습니다. 지금부터는 직주융합의 담론들이 생겨날 것이라 생각합니다. 리모트 워크도 정착되어가고 있는 상황입니다만, 전혀 출근을 하지 않고도 해결되는 업무는 언젠가 AI로 대체될지도 모릅니다. 그런 의미에서는 사람들이 모여서 지적 생산을 하는 장소의 가치가 높아질 수도 있습니다. 누구든지 자유로이 출입하고 의견교환이 가능한 장소가 퍼블릭 스페이스이며, 그러한 장소가 도심뿐만 아니라 주택지에 만들어지는 것도 있을 수 있습니다.

— 앞으로 퍼블릭 스페이스는 누가 어떻게 만들고, 어떻게 매니지먼트되어야 합니까?

데구치: 현재 대학 수업을 전부 온라인으로 대체하고 있습니다. 지금까지의 수업에서는 수업 후 학생으로부터 질문을 받는다든지 교실에 남은 학생들과 토론을 하는 경우가 많았습니다. 온라인에서는 스위치를 누르는 순간 수업이 끝나버립니다. 또 텔레워크를 계속하고 있자면, 지금까지 이동하는 시간이 머리를 전환하는 유익한 시간이었다는 것을 새삼 깨닫습니다. 업무 중간 휴식을 취하거나 잡담을 하는 장소로써의 공공공간과 그 곳에서의 행위를 즐기는 시간은 일상생활에 필요하다고 느낍니다.

팬데믹 상황에서 공원의 놀이기구에 자물쇠가 채워져 사용금지된 광경을 목격했습니다만, 이러한 상황 속에서 퍼블릭 스페이스의 관리는 어렵다고 생각됩니다. 행정기관에 관리를 맡기면 리스크를 피하려고만 하기 때문에 거친 표현으로 말하면 사고정지(思考停止) 상태가 됩니다.

그렇다고 이용자의 매너에 맡기는 것도 한계가 있습니다. 이런 상황 속에서 하나의 방법으로 데이터를 이용해 매니지먼트 방법이 가능할 것입니다. 광장 등의 이용 상황을 데이터화해 스마트폰을 통

Kishii: First, there are different categories of things to consider, depending on whether we're talking about the period "during the pandemic " or "after the pandemic." The former is expected to be short in duration, so perhaps we should focus our discussion on the latter. We now know that COVID-19 infections can spread from person to person in two ways: by droplet transmission and contact transmission. We've therefore taken measures to restrict people's activities and contacts. Though telecommuting and online classes have been widely adopted, we've found it difficult to entirely eliminate person-to-person contact, despite the numbers of people who are infected. I've come to understand the limitations of remote activities and to appreciate the value of meeting and exchanging information with people in person. These issues will affect our lifestyles considerably in the future.

Takami: In the past, there was nothing wrong with urban activities that brought crowds of people into close contact with each other. In the future, however, we will no doubt think carefully about crowd density in a variety of settings. The pandemic is also likely to increase the value of provincial areas, which could help mitigate overconcentration in Central Tokyo and lead to an urban structure of more homogenous density. In the past, public spaces in the city center mostly served as places for events and relaxation. In the future, they will no doubt serve a greater variety of purposes—telecommuters, for example, might spend hours working in them every day. As for suburban parks, these were designed as places for relaxing and playing sports on weekends and holidays. In the future, people might start using them on a more daily basis.

Deguchi: People are spending more time working from home due to the widespread adoption of telecommuting. When I stroll in my neighborhood and stop by the local park, I see lots of different people relaxing there: workers tired of telecommuting and seeking diversion, parents playing with their children, and so on. In public spaces, people ought to be guaranteed the freedom to spend their time as they please—be it reading books, losing themselves in their thoughts, or whatever—even though others are also using the spaces. In our last round-table discussion, I said that today "people expect contemporary open spaces in cities to draw crowds." Well, I think people have become too fixated on the idea of public spaces as devices for drawing crowds. I agree with what Mr. Takami said: that the spaces may come to figure more prominently in people's daily lives by functioning like parks in residential neighborhoods.

Nakai: I too have been telecommuting during the pandemic. This has caused me to reflect on the fact that the contemporary way of living and working—of residing in the suburbs and commuting to one's workplace—is something that only came about in the past 100 to 150 years. From the end of the Edo period [(1603–1868)] through the Taisho period [(1912–1926)], Japan experienced numerous epidemics of cholera, typhoid fever, and tuberculosis. These diseases, transmitted through well water, caused many deaths. COVID-19, in contrast, is most likely to be transmitted in tightly sealed spaces. This is ironic because cities, from the modern era up to the present, have developed as a result of people dividing up and sealing off spaces, then exercising precise control over the functions and environments in the resulting spaces. In the post-COVID era, the focus of human efforts to control may transition to cyberspace—to virtual space. But even if this is so, our appreciation for basking in sunlight, breathing fresh air, and having face-to-face conversations with other people is unlikely to change. The true value of public spaces is that they facilitate these sorts of behaviors. One could also say they reflect our attempts to achieve great freedom in real space. The COVID-19 pandemic could be an opportunity in disguise if it relativizes the modes of working and living we've taken for granted and reveals to us new possibilities for freer cities and public spaces.

Kishii: The advent of railways allowed people to leave deteriorating environments in the centers of cities and build houses in suburban residential districts. This compartmentalization of people's lives was based on the concept of shokuju bunri [("separated working and living")] that was promoted [in Japan] at the time. However, the development of advanced information networks led to the emergence of a new world that rendered the notion of shokuju bunri obsolete. In the future, people are more likely to talk of shokuju yugo [("integrated working and living")]. Now telecommuting has taken root, but the types of work that can be performed remotely, without ever going in to an office, may soon be performed by artificial intelligences. In light of this, there's a good chance that a higher value will once again be assigned to places where people can congregate and engage in intellectual production. Places where everyone is free to come and exchange ideas are none other than public spaces. I predict that such spaces will increasingly be built not only in city centers but also in residential districts.

—**Going forward, who should be responsible for public spaces? How should the spaces be managed?**

Deguchi: Right now, I'm teaching all of my college courses online. In the past, I would often stick around after class to take questions and talk with students. Online classes, however, are over with the flip of a switch. One thing I've learned during this prolonged period of telecommuting is that my commuting time provided me with a valuable chance to reset my mind. The coronavirus has helped me realize the importance of the most resonant times and places in my daily life. I now know that I need breaks between my different tasks. I also need to spend time in public spaces, to communicate with other people and enjoy the activities that take place there.

After the state of emergency was declared, I saw the playground equipment in a park being taped up, to prevent it from being used. Public spaces are no doubt difficult to manage during the pandemic. When the government is tasked with this management, it tends to do it in a mindless way, since it strives to minimize risk. And of course there are limits to leaving everything up to the manners of the people who use the spaces. One way we might deal with this is by implementing data-based management methods. By converting the usage of plazas and so on into data and making the data available through smartphones in realtime, for example, we might bring about behavioral changes in people who use the spaces. We might implement management methods that meld cyberspace, or virtual space, with physical space, or real space, as is extolled in Society 5.0*3 When it comes to management methods relying on data—for example, installing cameras in parks or on sidewalks while taking necessary precautions with personal data— at present, this data is highly secured. Few employees and organizations have access to it and there is resistance to letting private corporations manage it. I think that organizations selected by the community ought to manage the data as well as the public spaces and facilities, with ethical oversight from third-party organizations. We've now entered a period of transition in the use of this sort of data. If we learn how to use it effectively we will be able to manage more intelligently the next time we find ourselves in these circumstances.

Nakai: You could say that in the past, people's experiences of working

해 리얼타임으로 확인가능하면 이용자의 행동 변용을 촉진할 수 있습니다. 「Society 5.0」에서도 주장하고 있듯, 사이버 공간과 현실공간을 융합시키는 것으로 매니지먼트해가는 것이 가능할 것입니다. 데이터를 활용해서 관리하는 경우, 예를 들면 공원이나 보도에 카메라를 설치해 개인정보 보호를 충분히 배려하면서 관리하는 방법을 생각할 수 있으나, 실제로는 데이터를 고도의 세큐리티로 관리하며 활용가능한 조직이나 인재가 아직 많지 않습니다. 외부의 민간기업에게 데이터 관리를 일임하는 것이 또 다른 리스크이기도 하므로, 가능한 지역에 신뢰가 있는 조직이 관리하면서 공공공간이나 공공시설을 매니지먼트해갈 수 있으면 어떨까 생각합니다. 그 경우 제3자 기관에 의한 윤리적 감시가 필요로 할 것입니다. 현재는 그러한 데이터 활용의 과도기입니다만 데이터를 유효하게 사용할 수 있게 된다면 지금과 같은 상황에 다시 처한다고 해도 현명하게 매니지먼트 가능하다고 생각합니다.

나카이: 지금까지는 사람이 공간을 관리하는 것을 통해, 일하고 생활하고 있다는 느낌을 실감했다고 말할 수 있습니다. 텔레워크가 널리 퍼져 일반화되어가면서 공간과 시간의 관리는 어느 정도 자유로워질 수 있을 것입니다. 그때 우리들은 무엇에 의지해서 삶의 실감과 보람을 얻을지 생각해보아야 할 것입니다. 어떤 시대도 실질적인 땅에 발을 딛고 있는 자신들의 장소, 사람들과의 관계, 커뮤니티는 필요할 것입니다. 사람이 태어나는 장소는 선택할 수 없으나, 어느 토지에서 생활하고 공부하고 아이를 키우고 삶을 마감할지 스스로 선택하는 것은 지금 이상으로 인간 삶의 충실도에 영향을 미칠 수 있습니다.

다카미: 본지에서 다루고 있는 퍼블릭 스페이스는 민간개발에 의한 것이 대부분입니다. 그 자체는 이익을 창출하지 않지만, 근방의 임대 공간의 부가가치를 부여하는 것으로 결과적 이익을 얻는 구조입니다. 그러나 차후로는 개발 주체가 작아질 수도 있습니다. 종래의 자본 논리로 관리하고 있던 장소가 관리가 불가능해지는 상황이 발생할 수도 있습니다.

나카이: 역사적으로 퍼블릭 스페이스의 역할을 담당했던 신사(神社)의 경내 등은 지역 고유의 방식으로 관리되었습니다. 그것이 근대가 되면서 중앙집권에 의해 공간이 존재하는 방식을 체계적으로 관리하는 사회로 변화해왔습니다. 그것이 지금 다시 진화를 필요로 하는 시기일지도 모르겠습니다. 다음 시대의 퍼블릭 스페이스는 새로운 생활방식에 맞춰서 시행착오를 거치면서 만들어가는 수밖에 없을 것입니다. 일반적으로 이노베이션은 보수적인 중핵에서 일어나는 것이 아니라, 다급하고 절실한 주변부의 현장에서 일어나서 번져 나간다고 생각합니다. 지금은 다음 시대의 퍼블릭 스페이스의 가능성을 좁히고, 한정하지 않도록 다양한 개개 현장의 시행착오를 살펴보며 유연하게 대응해가는 것이 필요합니다. 그때 퍼블릭 스페이스가 담당하는 「지역」이라 함은 무엇인가, 동네의 커뮤니티의 스케일인지 아니면 기초자치체와 같은 큰 스케일인지, 재정의하는 것도 필요로 할 것입니다.

기시이: 행정기관은 특정의 누군가를 위해서 무언가 특별한 조치를 취하려 하지 않기 때문에 근대의 행정에 의해 관리되는 퍼블릭 스페이스에서는 전체 이용자를 「평등」하게 취급해 왔습니다. 그러한 퍼블릭 스페이스를 관리해 가기 위해서는 지역 매니지먼트의 수법이 적합하다고 생각합니다만, 향후 그 관리방식이 변화할지도 모르겠습니다.

— 지금부터의 퍼블릭 스페이스에 무엇을 기대하겠습니까?

나카이: 현대사회에 있어서 개인은 시스테믹하게 관리되는 대신 그 범위에 있어서 자유와 평등을 획득하고 있다고 말할 수 있습니다. 그것은 반드시 부정적으로 볼 것이 아닙니다만, 그것만으로는 새로운 가치가 생성되지 않는다는 것이 문제입니다. 점점 변화하는 시대에 적응하기 위해서는 유연하게 새로운 가치를 생산하는 사회적 공간이 필요합니다. 그를 위해서는 다양한 사람들이 직접 만나고 대화하고 교류하면서 새로운 가치를 싹 틔우고 양성하는 장소가 필요하고, 현대에 있어서 퍼블릭 스페이스는 그러한 역할을 담당하는 하나의 배양기가 될 수 있다고 생각합니다. 그러한 퍼블릭 스페이스를 공공도 민간도 관계없이 사회로서 셰어가능하다면 이상적일 것입니다.

데구치: 공공공간은 각각 계획된 시대의 사고방식과 가치관이 나타나는 공간이고, 도시의 특징과 시대를 보는 렌즈라고도 할 수 있습니다. 독자들에게는 책을 통해 퍼블릭 스페이스 자체의 디자인을 생각해 봄과 동시에, 그 속에 잠재되어 있는 구조나 사고, 각 시대의 제도와 정책도 함께 읽어내길 기대합니다.

다카미: 지금까지의 공공공간을 자유도가 높은 관리 시스템에 의해 만들어온 한편, 이용자나 그곳에서 실제로 일어나는 행동의 다양성을 두고 상대적으로 생각해보면 한정적인 공공공간이 많았다고 느낍니다. 그러한 한정과 한계를 점점 넓혀가는 것을 기대해봅니다.

기시이: 사람들이 만나는 장소와 다양한 활동을 가능하게 하는 장소를 다같이 만들고 관리하는 것이 중요합니다. 특히 도시 간 경쟁이 치열한 현대에 장소의 매력을 만들어내기 위해서는 필수적입니다. 종래와 같이 행정기관이 관리하는 퍼블릭 스페이스는 한계가 있고, 한편으로 이윤을 추구하는 민간이 만드는 독자적인 오픈 스페이스 또한 한계가 있습니다. 그 어느 쪽도 해당되지 않고, 어느 쪽도 해당되는 공공공간을 어떻게 해서 도시와 지역사회에 만드는 것이 가능할지가 기대되고 있다고 생각합니다.

본지는 최근 관심이 높아지고 있는 퍼블릭 스페이스에 대해, 최근에 도쿄에 만들어진 것을 세계를 향해 발신하고자 기획되었습니다. 독자들에게는 도쿄에 다양한 퍼블릭 스페이스가 존재하는 것과 그것들이 생성되는 배경을 이해하는 것과 동시에, 현지에 가서 그 실공간을 체험하고 확인하길 바랍니다. 인프라스트럭처로서 가치 있는 퍼블릭 스페이스를 만들기 위해서, 또 지금 있는 것을 유지관리하기 위해서도 다같이 셰어해 가는 것이 중요하다고 생각합니다. 그것이 콜라보레이티브 디자인이라고 불리고, 그렇게 해서 만들어진 장소가 뉴 어반 인프라스트럭처일 것입니다.

(2020년 7월 2일, Web회의   정리: 본지편집부)

and living were mediated by their subjection to spatial management. The universalization of telecommuting, however, has enabled people to live with a new degree of freedom from both spatial and temporal management. Now, as we seek meaning in our lives, we must consider what will serve as the new foundations. I am of the opinion that people in any era need to have their own places, interpersonal connections, and communities rooted in physical places. Although people can't choose where they are born, they can choose where they live, study, raise children, and die. These choices may have a greater effect on the meaningfulness of people's lives than they did in the past.

Takami: Most of the public spaces in this magazine were realized through private development projects. The spaces themselves might not generate profits but they add value to nearby rental spaces, and so form part of a system that generates profits. However, we may find that there is less development in the future. We may also find that some spaces that were managed according to conventional, capitalistic logic are no longer manageable.

Nakai: The precincts of shrines and other spaces that once functioned as public spaces were managed using systems rooted in their localities. The advent of the modern era, however, transformed our society into one in which spaces are managed uniformly and systematically by centralized authorities. Our society might now be undergoing another transformation. I suppose there will be no choice but to create the public spaces of the next era through a process of trial and error—to reestablish ways of living and working rooted in our specific localities. In general, innovation occurs and spreads not from the conservative center of societies but from their edges, where a greater sense of urgency tends to prevail. To avoid limiting the possibilities of future public spaces, we must respond flexibly while encouraging trial and error at various sites. In the process, we must ask ourselves which "local communities" will bear responsibility for the spaces. Will it be handled at the scale of chonaikai [("neighborhood associations")] or at that of municipal governments? Will we have to devise new solutions?

Kishii: The government disapproves of taking special steps to benefit any particular parties. It gives equal treatment to everyone who uses public spaces managed by the government. Today, however, people are beginning to discover new values in public spaces that could once only be found in private spaces. In the future, we will seek new balance between "cooperation based on equality" and "individual freedoms." While I believe that "area management" methods relying on local communities will be suitable for managing these spaces, I recognize that management practices may change in the future.

—What are your hopes for public spaces in the future?

Nakai: In contemporary society, individuals aren't managed systematically; they experience freedom and equality, within certain limits. While this is nothing to scoff at, the issue is that this alone does not generate new values. In order to confront the rapidly changing world, we need social spaces that are capable of flexibly generating new values: places where various types of people come into contact with each other, talk, exchange ideas, and confront each other. We need places where new values can sprout and ferment. Public spaces can serve as incubators that fulfill this role. I hope these public spaces will be shared parts of our society, whether they are in fact public or private.

Deguchi: Public spaces express the thoughts and values of the eras in which they were designed. They can also be described as lenses for viewing the different eras and characteristics of cities. I hope that readers will take pleasure in the designs of the public spaces presented in this magazine. I also hope that they will come to understand the structures and thinking behind the designs as well as the systems and policies of the different eras.

나카이 유. Mr. Yu Nakai.

Takami: Public spaces built in the past have management systems that provide a high degree of freedom. Nevertheless, when I consider the lack of diversity among the people who use the spaces, and the narrow range of behaviors that occur in them, I must conclude that many of the spaces are in fact rather limiting. I hope these limitations will fall away in the future.

Kishii: It is important that everyone have a hand in creating and managing the places where people meet and that these places permit a variety of activities. It's particularly necessary today, when people are extolling the virtues of interurban competition, because people's participation is required to create charming places. We will not succeed in achieving such participation by relying solely on conventional public spaces that are managed by the government. Having said that, there are also limits to the kinds of open spaces that private companies can create. I hope that cities and regional communities will somehow figure out how to create public spaces that are both public and private and that are neither public nor private.

Public spaces have been attracting more interest recently. Our planning for this issue began with a desire to share the recently created public spaces in Tokyo with the rest of the world. We hope to inform readers about the variety of Tokyo's public spaces and the contexts in which they came about. Additionally, we hope readers will visit their sites, to examine and experience their actual spaces. If we are to create public spaces of value, spaces that truly serve as urban infrastructure, and if we are also to manage and maintain our existing public spaces, the spaces must be shared by everyone. This is what we hope to communicate with the phrase "cooperative design." Places created in this way will serve as our "new urban infrastructure."

(July 2, 2020, by web conference;
      edited by Shinkenchiku-sha Editorial Department)

*1 The Specific Block System is a system that allows floor area ratios and heights that differ from those established by Japan's Building Standards Act to be established for blocks in districts or zones that are subject to Japan's City Planning Act. The goal of the system is to maintain or improve existing urban areas, for the sake of city planning.
*2 The Comprehensive Design System is a system that relies on Japan's Building Standards Act to approve exceptional easing of city planning restrictions. Floor area ratios, absolute height restrictions, setback regulations, and so forth may be eased for plans deemed to contribute to the maintenance or improvement of urban environments by including public open spaces.
*3 Society 5.0 refers to human-centered societies that balance economic advancement with the resolution of social problems using systems that highly integrate cyberspace and physical space.

Hareza 이케부쿠로·나카이케부쿠로 공원의 지역 매니지먼트
Area management for Hareza Ikeburo/Naka-Ikebukuro Park **58**

IKEBUKURO

나카노 센트럴 파크 파크 에비뉴
나카노 시키노 모리 공원
Nakano Central Park　Park Avenue
Nakano Shiki no Mori Park **31**

NAKANO

게이오 플라자 호텔 Wood Promanade Garden
신주쿠 부도심 4호선 55HIROBA
Keio Plaza Hotel's Wood Promenade Garden
Shinjuku Fukutoshin Line 4　55HIROBA **32**

신주쿠 부도심 지역을 대상으로 한 환경개선위원회의 활동
Initiatives by Shinjuku West Area Development and Management Committee **50**

신주쿠역 Suica 펭귄 광장
Shinjuku Station Suica's Penguin Park **03**

SHINJUKU

키타아오야마 3초메 지역만들기 프로젝트
Kita-aoyama 3-chome District Project **37**

SHIBUYA CAST.의 공개 공지 활용
Utilization of privately owned public space at SHIBUYA CAST. **56**

시부야 구립 미야시타 공원
SHIBUYAKURITSU MIYASHITA PARK **45**

시부야 파르코·휴릭빌딩 나카시부 거리
Shibuya PARCO·HULIC building　Nakashibu Street **18**

시부야 파르코·휴릭빌딩 루프탑 파크 등
Shibuya PARCO·HULIC building　ROOFTOP PARK, etc. **44**

시부야역 중심지역 보행자 네트워크
Shibuya Station Central District　Pedestrian Network **13**

시부야역 앞 지역 매니지먼트 옥외광고물 실증 실험
Area management in front of Shibuya Station　Demonstration experiment of outdoor advertisements **57**

시부야역 동쪽 출구 지하광장
Shibuya Station East Exit Underground Plaza **42**

시부야 스트림 앞 이나리 다리 광장·킨노 다리 광장
Inari Bridge Square, Konno Bridge Square in front of SHIBUYA STREAM **41**

에비스 가든 플레이스
시계 광장·센터 광장·샤또 광장
YEBISU GARDEN PLACE
Marionette Clock Square, Central Square, Château Square **28**

SHIBUYA

무사시코야마역 더 플라자
Musashi-koyama Station　The Plaza **06**

후타고타마가와역 후타고타마가와라이즈 갤러리아
Futako-tamagawa Station　Futako Tamagawa Rise　Galleria **05**

FUTAKO-TAMAGAWA

도쿄 사잔 가든 고탄다 수변 만남의 광장
Tokyo Southern Garden　Gotanda Fureai Waterfront Plaza **35**

오사키역 주변지구 데크 네트워크
Area around Osaki Station　Deck Network **12**

Think Park : Think Park Forest
NBF 오사키 빌딩 녹지
ThinkPark　ThinkPark Forest
NBF Ohsaki Bldg.　Green Zone **29**

0　400　1000　2000M N

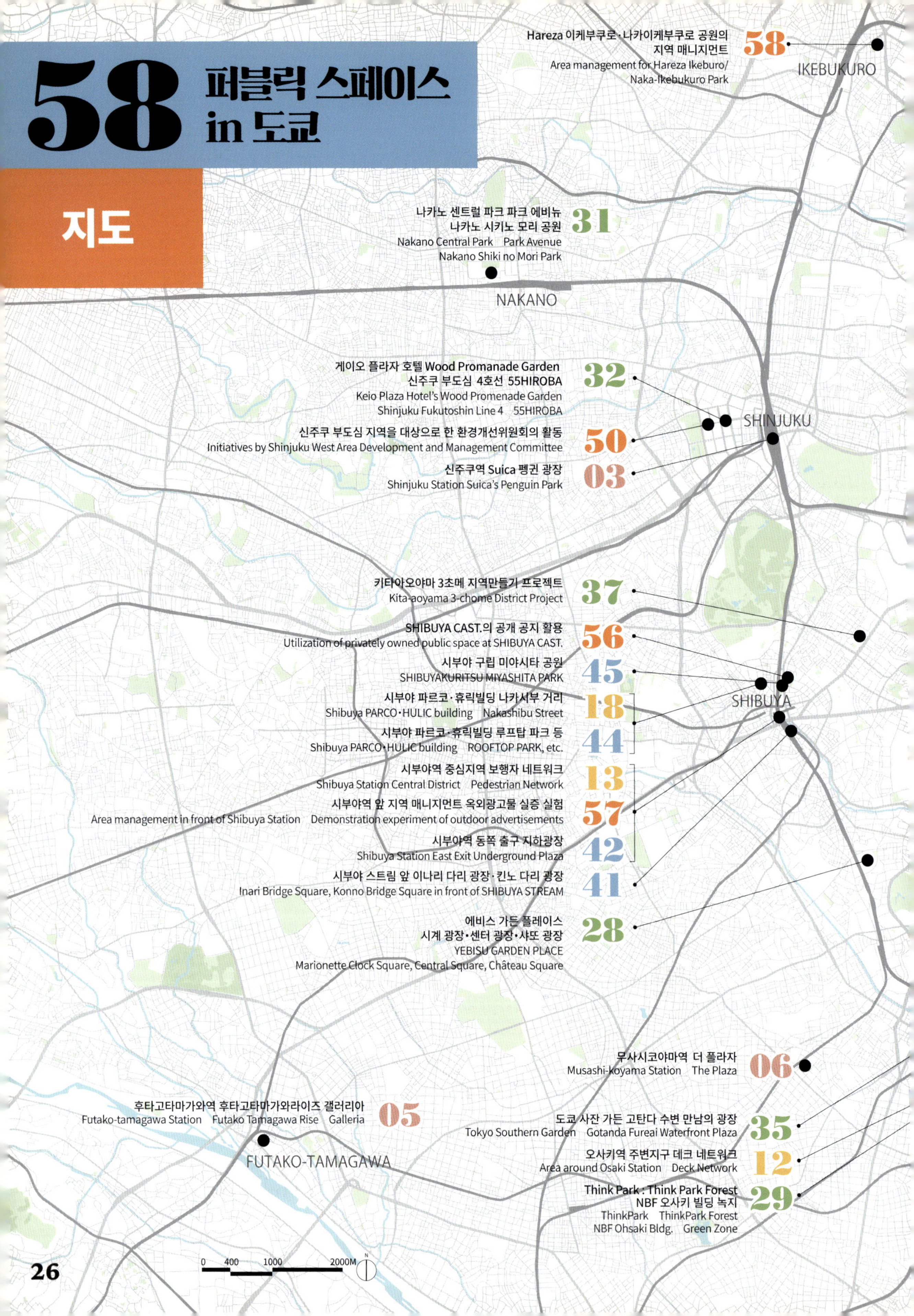

30 아이 가든 에어 히라카와 오솔길 기타
i-Garden Air Hirakawa no Michi, etc.
49 와(和) 테라스, 아와지 공원을 중심으로 한 고베아와지초 지역 매니지먼트
Kanda Awajicho area management centered around WATERRAS and Awaji Park
22 오테마치 플레이스
Otemachi Place
선큰 가든 센트럴 프롬나드 료칸 사쿠라 브릿지
Sunken Garden, Central Promenade, Ryukan-sakura Bridge
21 오테마치 타워 플라자
The Otemachi Tower Plaza
01 도쿄역 마루노우치 입구 역 앞 광장
Tokyo Station Marunouchi Station Plaza
TOKYO
33 카스미 테라스·케야키 광장
Kasumi Terrace, Keyaki Square
36 도쿄 스카이트리 타운® 스미다구립 오시나리 공원
TOKYO SKYTREE TOWN Oshinari Park
48 니혼바시 무로마치 지역 매니지먼트
Nihonbashi Muromachi area management
09 지하철 미츠코시마에역 니혼바시 지하 보도
Subway Mitsukoshimae Station Nihonbashi Underground Walkway
47 니혼바시 강가 퍼블릭 스페이스
Public space along Nihonbashi River
08 지하철 니혼바시역 지하 보도
Subway Nihombashi Station Underground Walkway
17 니혼바시갤러리아
Nihonbashi Galleria
02 도쿄역 야에스 입구 역 앞 광장
Tokyo Station Yaesu Station Plaza
14 마루노우치 나카도오리
Marunouchi Naka-dori Street
20 다이마루유 지구 지하 보행자 네트워크
Otemachi-Marunouchi-Yurakucho District Underground Pedestrian Netwo
52 다이마루유 지구 지역 매니지먼트
Area management of Otemachi-Marunouchi-Yurakucho District
11 도쿄역 앞 지구 버스 터미널
Tokyo Station Regional Bus Terminal
34 오가와바타 리버시티 21 츠쿠바 공원 이시카와지마 공원
Okawabata River City 21 Tsukuda Park, Ishikawajima Park
19 도쿄 스퀘어 가든 공공 통로 교바시 에도그랑 공공 통로
TOKYO SQUARE GARDEN Through Passage KYOBASHI EDOGRAND Through Passage
04 유락초역 앞 광장 유락초이토시아
Yurakucho Station Plaza Yurakucho Itocia
07 도쿄 미드타운 히비야 히비야 아케이드
TOKYO MIDTOWN HIBIYA Hibiya Arcade
15 도쿄 미드타운 히비야 히비야 스텝 광장 등
TOKYO MIDTOWN HIBIYA Hibiya Step Square etc.
43 도쿄 미드타운 히비야 파크뷰 가든
TOKYO MIDTOWN HIBIYA Park View Garden
51 히비야 스텝 광장 및 주변지구의 지역 매니지먼트
Area management of Hibiya Step Square and the surrounding area
23 시오도메 시오사이트 지하 보도와 선큰 광장
Shiodome Sio-site Underground Walkway and Sunken Plaza
53 신토라노몬의 스트리트 활용
Street utilization of Shintora-dori Avenue
38 토라노몬힐스 오발 광장
Toranomon Hills Oval Plaza
25 토라노몬힐스역 주변 네트워크
Network around Toranomon Hills Station
16 아카사카·토라노몬 녹지대 구상 아카사카 인터시티 AIR 보도상 공지 등
Akasaka-Toranomon Green Corridor Project Akasaka Intercity AIR Privately Owned Sidewalk Open Space, etc.
24 롯본기 1초메역 주변 네트워크
Roppongi 1-chome Station Network
26 도쿄 미드타운 미드타운·가든 히노키초 공원
TOKYO MIDTOWN MIDTOWN GARDEN Hinokicho Park
55 도쿄 미드타운 지역 매니지먼트
Area management of TOKYO MIDTOWN
39 롯본기힐스 66 플라자 Metro Hat
Roppongi Hills 66 Plaza, Metro Hat
54 롯본기힐스 타운 매니지먼트
Roppongi Hills town management
SHINAGAWA
40 시나가와 시즌 테라스 이벤트 광장 시바우라 중앙 공원
Shinagawa Season Terrace Event Square Shibaura Chuo Park
10 다카나와 게이트웨이역 보행자 광장
Takanawa Gateway Station Pedestrian Plaza
27 시나가와 센트럴 가든
Shinagawa Central Garden
46 국도 15호 시나가와역·서쪽 출구 역 앞 광장
National Route 15 / Shinagawa Station West Square
OSAKI
00 MOBILITY HUB
00 SEAMLESS LINKAGE
00 URBAN GREEN CORE
00 MULTI-LAYERED DEVELOPMENT
00 AREA MANAGEMENT

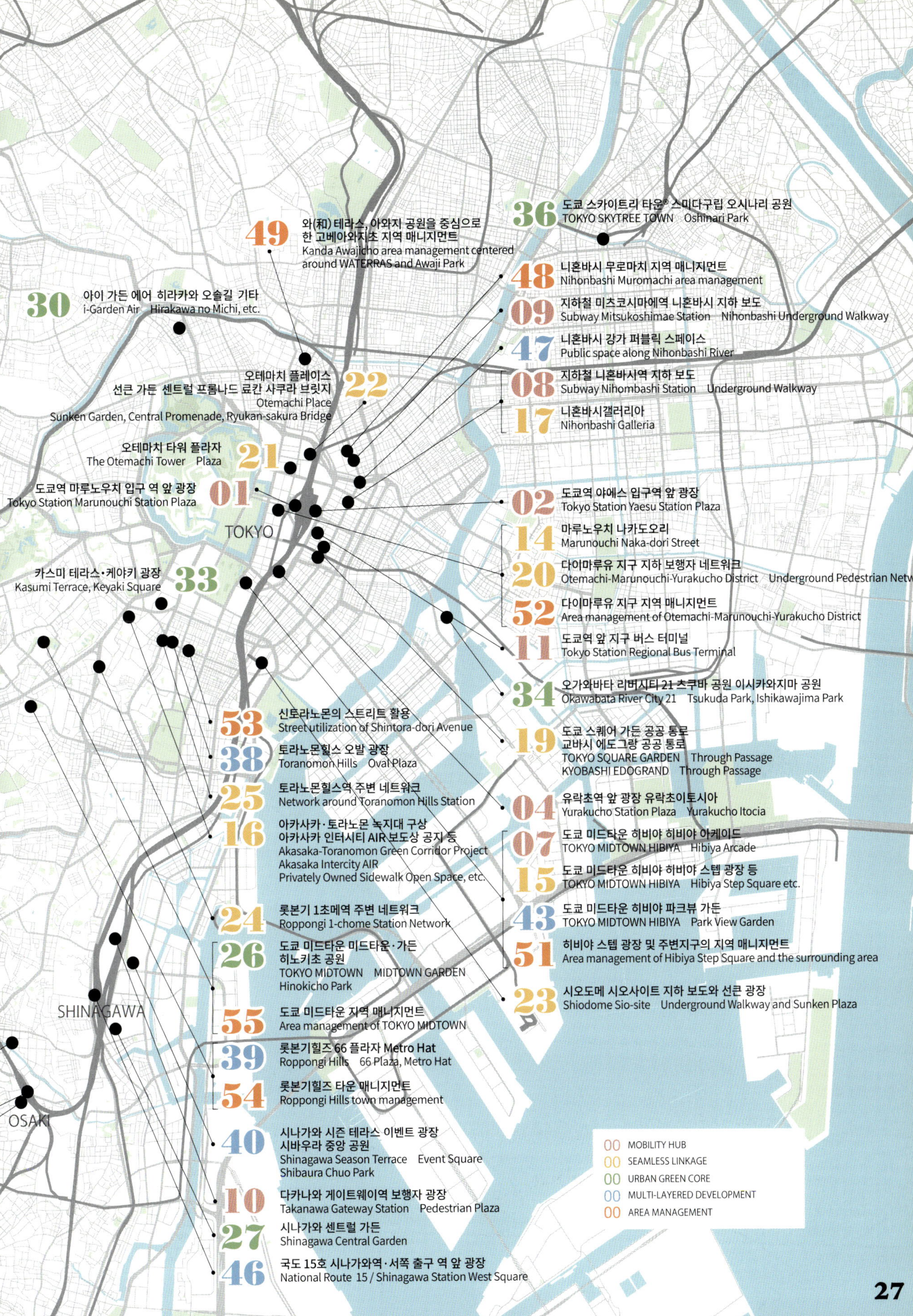

27

BUS
PM
TAXI

MOBILITY HUB

Public spaces with access
to transportation

모빌리티 허브가 되는
퍼블릭 스페이스

MAP p.189

MOBILITY HUB

북측에서 본 관경. 교통광장을 재편해서 만들어진 마루노우치 중앙공원
View from the north. Marunouchi Central Plaza which was reorganized from the traffic square.

도쿄의 상징으로 걸맞은 역 앞 광장의 재편

도쿄역 마루노우치 출구역 앞 광장(2017)

도쿄역 마루노우치 역사의 보존·복원에 맞춘 재정비 구상은 역사·역 앞 광장·교코도오리 등 마루노우치 출구 주변 도로까지 포함한 지역 전체의 토털 디자인으로 전개되었다. 역 앞 광장은 보행자 공간을 중심으로 한 상징적인 퍼블릭 스페이스로 다시 태어났으며, 복원된 역사와 함께 수도 도쿄의 얼굴을 만들고 있다.

Station plaza reorganization to represent the capital city of Tokyo
— Tokyo Station Marunouchi Station Plaza

01

Station front as a pedestrian prioritized plaza

역의 정면을 보행자 전용의 광장으로

정부, 대학, 기업으로 조직된 「도쿄역 주변 재생정비 연구위원회」는 정비 방침을 만들고, 도시계획도로와 교통광장, 보행자 전용도로 등에 관한 도시계획의 결정/변경을 시행하였다. 이에 따라 역 앞의 대부분을 차지하던 자동차 교통은 남북으로 집약시켜 중앙에는 교코도오리와 연결된 약 6500㎡의 대규모 보행공간인 「마루노우치 중앙광장」이 생겨나게 되었다.

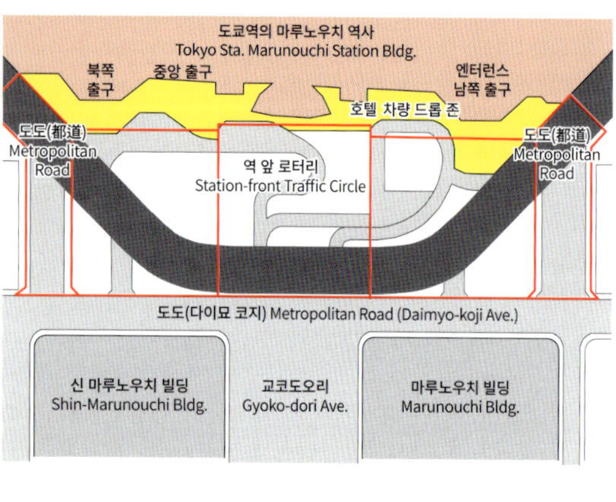

개발 전 배치　Site, before the development

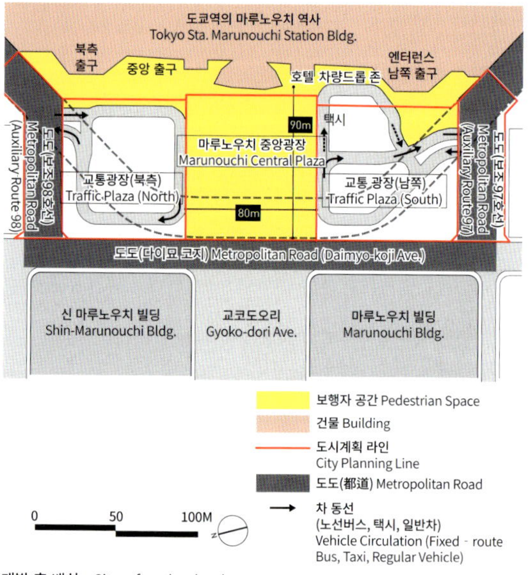

개발 후 배치　Site, after the development

남측에서 본 관경. 마루노우치 중앙광장은 남과 북의 교통광장과 중앙의 대규모 보행 공간으로써 기능을 함.
View from the south. Marunouchi Central Plaza functions as traffic plazas on the south and north and a large pedestrian space in the center.

개발 전의 도쿄역 마루노우치 로터리.
Tokyo Station Marunouchi traffic circle before the development.

MAP p.189

02

MOBILITY HUB

그랜드 루프와 녹지로 둘러싸인 도쿄역 야에스 출구 역 앞 광장
Tokyo Station Yaesu Station Plaza with its large roof, surrounded by greenery.

그랜드 루프와 녹지로 둘러싸인 역 앞 광장

도쿄역 야에스 출구 역 앞 광장(2014)

역 앞 광장과 소유자가 다른 3곳의 건축 부지를 일체적으로 계획하여 상징적인 그랜드 루프와 풍부한 녹지와 함께 대규모의 보행자 공간을 동시에 실현하였다. 이것으로 역 앞 광장의 다양한 액티비티와 교통광장 기능을 그랜드 루프와 녹지가 부드럽게 둘러싼 공간으로 도쿄역 야에스 출구 광장이 새롭게 태어났다.

북측에서 본 관경. 개발 전의 도쿄역 야에스 출구(2001년경)
View from the north. Tokyo Station Yaesu Entrance before the development (c. 2001).

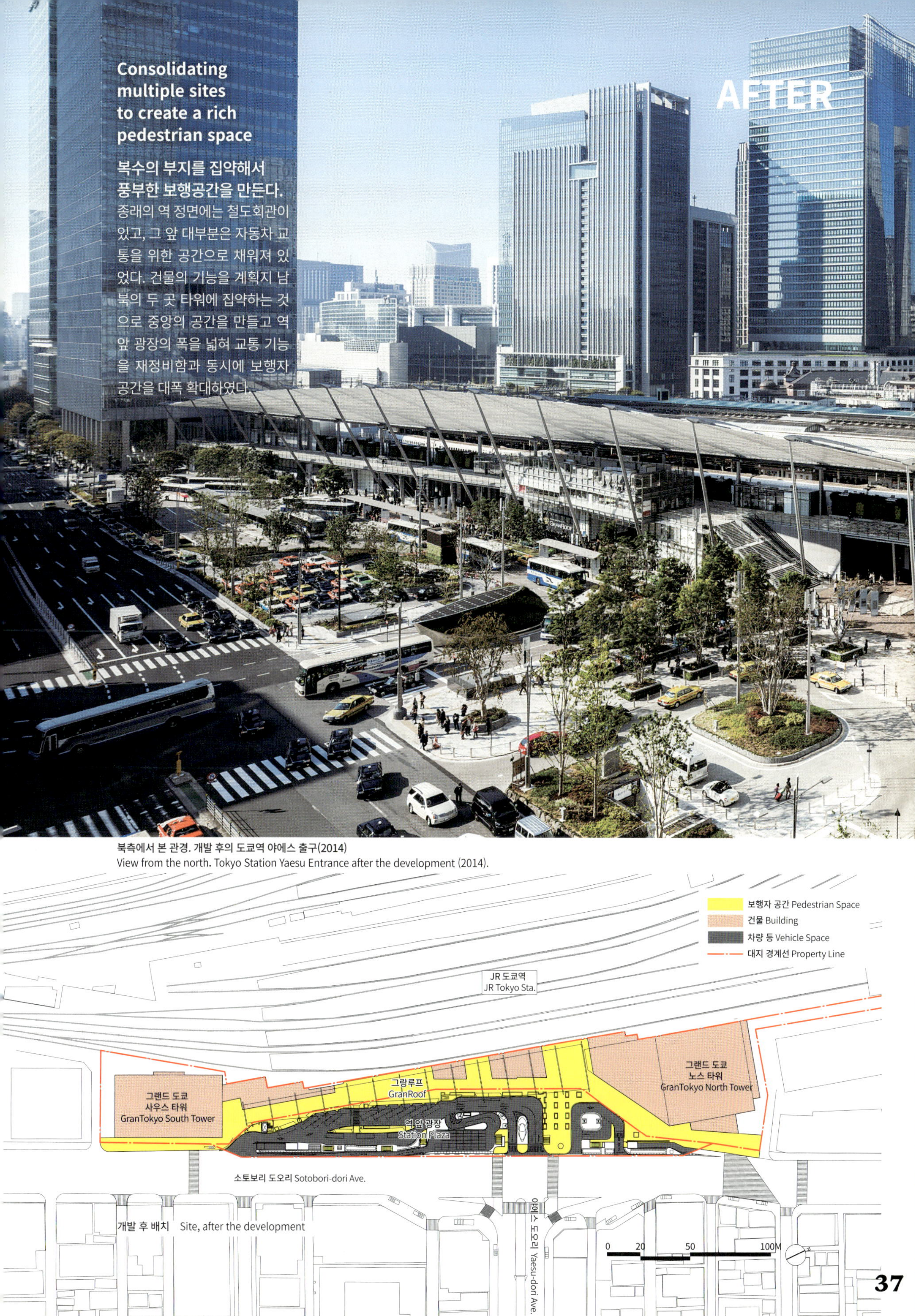

Consolidating multiple sites to create a rich pedestrian space

복수의 부지를 집약해서 풍부한 보행공간을 만든다. 종래의 역 정면에는 철도회관이 있고, 그 앞 대부분은 자동차 교통을 위한 공간으로 채워져 있었다. 건물의 기능을 계획지 남북의 두 곳 타워에 집약하는 것으로 중앙의 공간을 만들고 역 앞 광장의 폭을 넓혀 교통 기능을 재정비함과 동시에 보행자 공간을 대폭 확대하였다.

북측에서 본 관경. 개발 후의 도쿄역 야에스 출구(2014)
View from the north. Tokyo Station Yaesu Entrance after the development (2014).

개발 후 배치   Site, after the development

Plaza over the rail tracks looking out over the trains
—— Shinjuku Station　Suica' s Penguin Park

전철이 내려다보이는 선로 위의 광장

신주쿠역 Suica 펭귄광장(2016)

신주쿠역은 역으로써 필요한 보행자의 체류 공간의 부재, 버스터미널을 포함한 교통 인프라의 포화상태 등 여러 문제를 안고 있었다. 그것들을 일체적으로 해결하기 위해 정비된 것이 남측 출구 선로 상공의 인공지반이다. 그 위에는 역 시설, 상업/문화 시설, 택시 승하차장, 고속버스터미널이 입체적으로 정비되었다. 2층에 마련된 주로 보행자를 위한 Suica 펭귄광장은 신주쿠역에서 발착하는 전철의 움직임을 내려다볼 수 있는 인기 스폿이 되었으며, 신미나미 개찰구와 연접한 사잔 테라스와 신주쿠 다카시마야를 데크로 연결시켜 보행자 네트워크의 기점이 되었다.

선로 위에 입체적인 기능이 집적된다.
Various functions are vertically layered on top of the railway tracks.

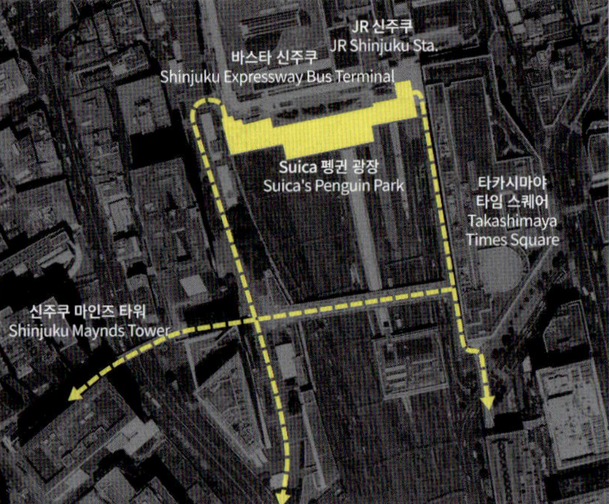

위 : Suica 펭귄광장(개업 당시)
아래 : Suica 펭귄광장은 보행자 네트워크의 기점이 된다.
Above: Suica's Penguin Park (at the time of opening).
Below: Suica's Penguin Park is a starting point of the pedestrian network.

Suica 펭귄광장을 내려다본 관경(개업 당시)
Looking down at Suica's Penguin Park (at the time of opening).

Complex railway network supporting the megacity Tokyo

메가시티 도쿄를 지탱하는 복잡한 철도망

아래는 세계 주요 도시의 인구밀도와 10km 권내 철도망을 표현한 것이다. 도쿄는 인구밀도 10,000 명/km²의 면적으로 다른 도시와 비교해 광범위하게 분포해 있는 세계 유수의 메가시티라 할 수 있다. 인구의 이동을 순조롭게 하기 위해 방사선의 철도망이 형성되어 있고, 중심부에는 다수의 노선이 복잡하게 얽혀 있으며, 하루 수백 만 명의 막대한 승하객 수가 있는 터미널역이 복수 존재한다. 앞에서 언급한 도쿄역이나 신주쿠역도 그 중 하나이며, 재개발의 가능성이 높은 역과 일체의 도시개발 속에서 양질의 퍼블릭 스페이스가 많이 생겨나고 있다.

TOKYO
도쿄

NEWYORK
뉴욕

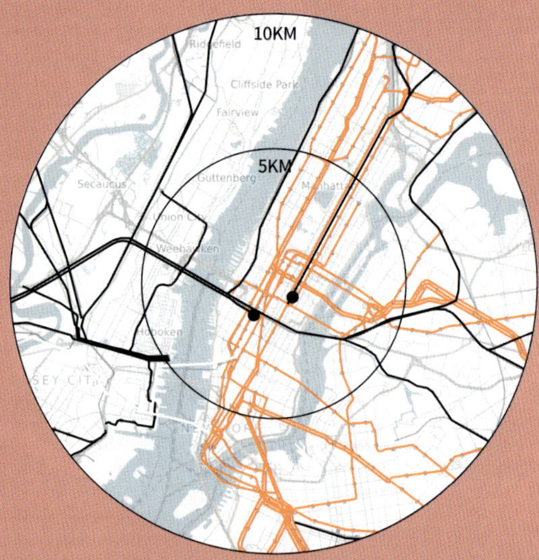

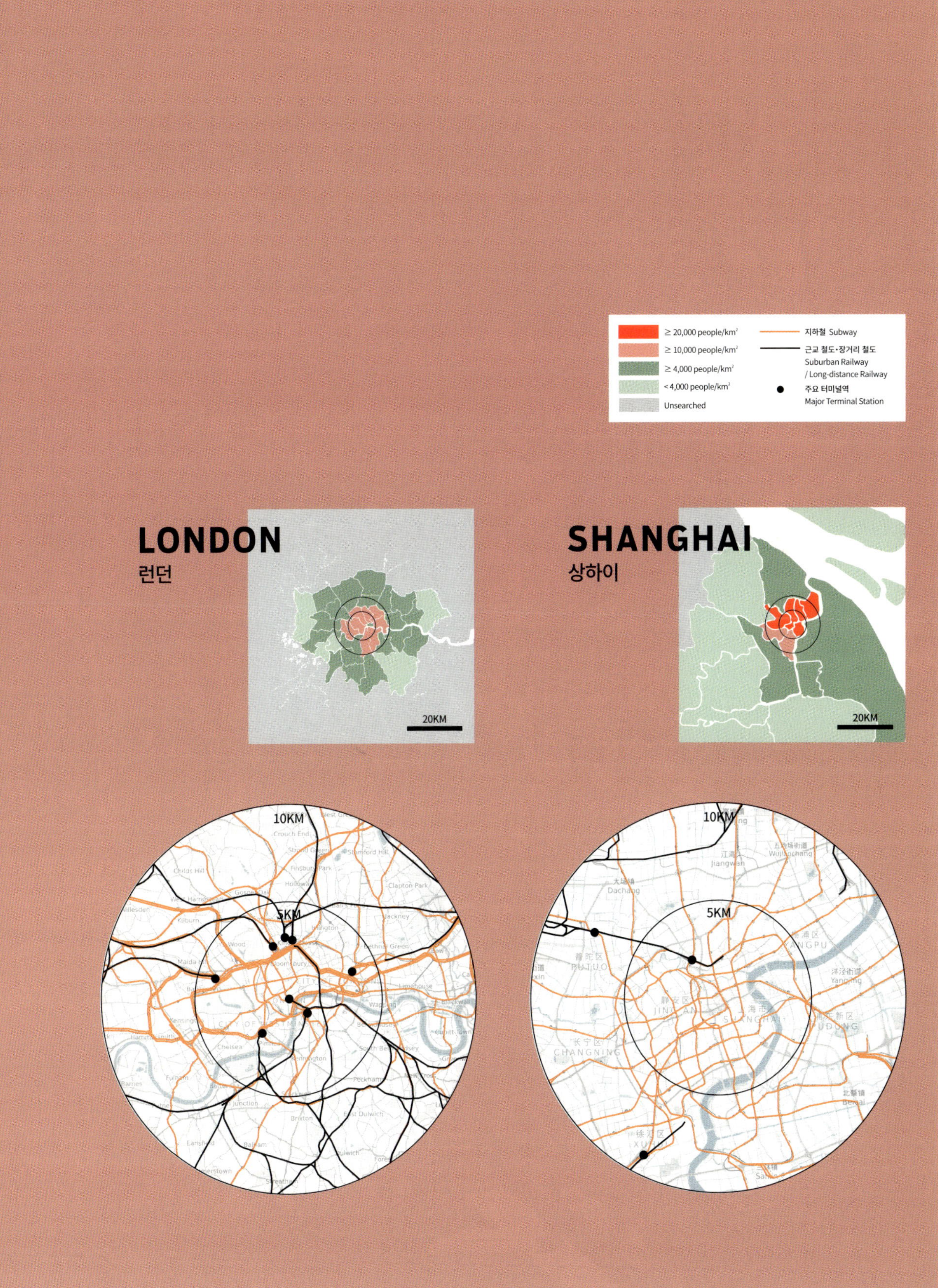

≥ 20,000 people/km²
≥ 10,000 people/km²
≥ 4,000 people/km²
< 4,000 people/km²
Unsearched
지하철 Subway
근교 철도·장거리 철도
Suburban Railway
/ Long-distance Railway
주요 터미널역
Major Terminal Station
LONDON
런던
20KM
SHANGHAI
상하이
20KM
10KM
5KM
10KM
5KM

BEFORE

개발 전의 유락초역 앞
View of the Yurakucho Station front before the development.

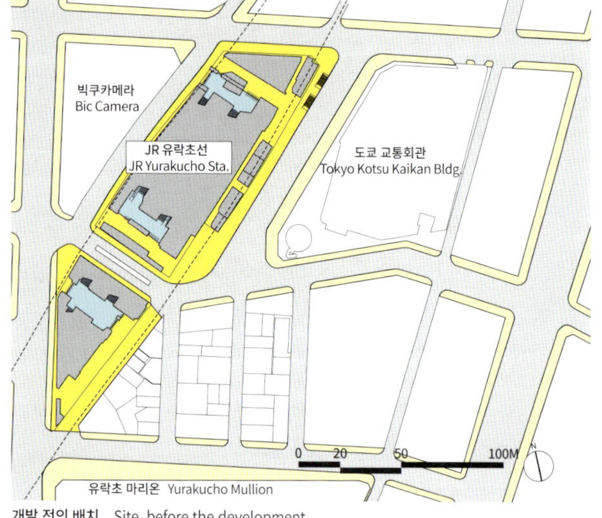

개발 전의 배치 Site, before the development

42

블록 통합으로 보행자 중심의 역 앞 광장을 확보

유락초역 앞 광장 유락초 이토시아 (2007)

시가지 개발에 의한 블록 통합으로 생겨난 도로 공간을 고가의 하부 도로와 함께 보행자 전용 도로로 만들고, 더불어 지하에도 교통광장을 정비하여 풍부한 보행자 공간이 역 앞에 생겨났다. 철도 사업자와 도로 관리자 등 소유자가 서로 다른 공간이면서도 통일된 디자인으로 일체화시킨 도시 공간이 되었다. 최근에는 약속 장소의 기능에 그치지 않고 지방의 특산물·각종 지역 활성화 이벤트가 개최되는 등 다양한 액티비티가 전개되고 있다. 더욱이 도시재생추진법인으로 지정된 「일반사단법인 유락초역 주변 도시만들기 협회」가 구의 협정을 준수하며 이 광장의 관리운영을 맡아 시행하고 있다.

Integration of city blocks to secure pedestrian-centric plaza in front of the station
— Yurakucho Station Plaza · Yurakucho Itocia

남서쪽에서 바라본 광경. 역 앞 광장에서는 다양한 액티비티가 전개된다.
View from the southwest. Various activities take place in the station plaza.

개발 후 유락초역 앞
View of Yurakucho Station after the development.

보행자 공간 Pedestrian Space
보행자 공간(지하) Pedestrian Space (Underground)
건물 Building
개찰 내 콘코스 Concouse, Paid Area
대지 내 경계선 Property Line
고가 라인 Elevated Railroad Line

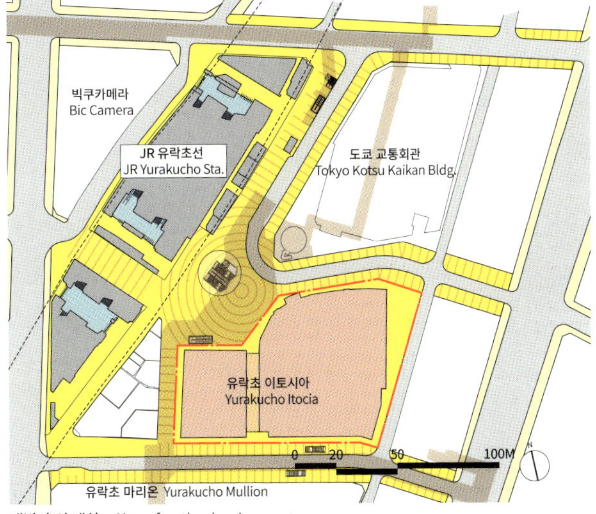

개발 후의 배치 Site, after the development

Pedestrian-centric development of station front through reconfiguration of the traffic plaza

— **Futako-tamagawa Station   Futako Tamagawa Rise   Galleria**

교통광장 재배치로
보행자를 위한 역 앞 공간 만들기

후타고타마가와역 후타고타마가와 라이즈 갤러리아(2011)

후타고타마가와역은 역 앞에 정비된 교통광장을 둘로 분리해 역에서 떨어진 위치에 재배치되어 있다. 이 방법은 역 앞을 보행자 중심의 광장으로 전환시킨다. 이렇게 해서 만들어진 역과 교통광장 사이에는 높이 30m의 개방적인 갤러리아와 상업시설이 계획되어 사람들이 오가며 다양한 이벤트가 열리는 도시의 리빙룸과 같은 장소가 되었다. 특히 갤러리아는 리본스트리트로 명명된 보행자 전용공간과 이어져 뒤쪽의 상업시설 및 문화시설과 일체적인 공간을 형성함과 동시에 부지의 서쪽 공원과 타마가와까지 연결된다.

후타고타마가와역의 퍼블릭 스페이스　Public space at Futako-tamagawa Station

왼쪽 위 : 정비 전의 후타고타마가와역 광장. 왼쪽 아래 : 리본스트리트는 후타고타마가와공원과 타마가와까지 연결된다. 오른쪽 : 후타고타마가와 라이즈 갤러리아
Above left: Futako-tamagawa Station. Plaza before the development. Below left: Ribbon Street leads to Futako Tamagawa Park and Tama River. Right: View of Galleria at Futako Tamagawa Rise.

The plaza connecting the station with the bustle of the town
—Musashi-koyama Station　The Plaza

역과 도시의 활기를 연결하는 광장
무사시코야마역 플라자(2019)

무사시코야마역은 역이나 상가의 이용자가 체류할 수 있는 거점인 광장과 그곳에서 아케이드 상가로 연결되는 관통 공간을 정비하고, 역 앞 광장에서 상점가로 이어지는 활력 공간을 정비하였다. 또한 상가에 접한 건물 저층부는 「도쿄의 매력 있는 도시 만들기 조례」에 근거한 「도시재생방침」을 정했으며, 재개발 등 촉진구역을 정하는 지구계획운용 기준에 따른 벽면의 위치에 의하지 않고 점포의 연속성을 고려한 계획으로 아케이드 상점가라고 하는 특징적인 퍼블릭 스페이스의 계승을 도모하고 있다.

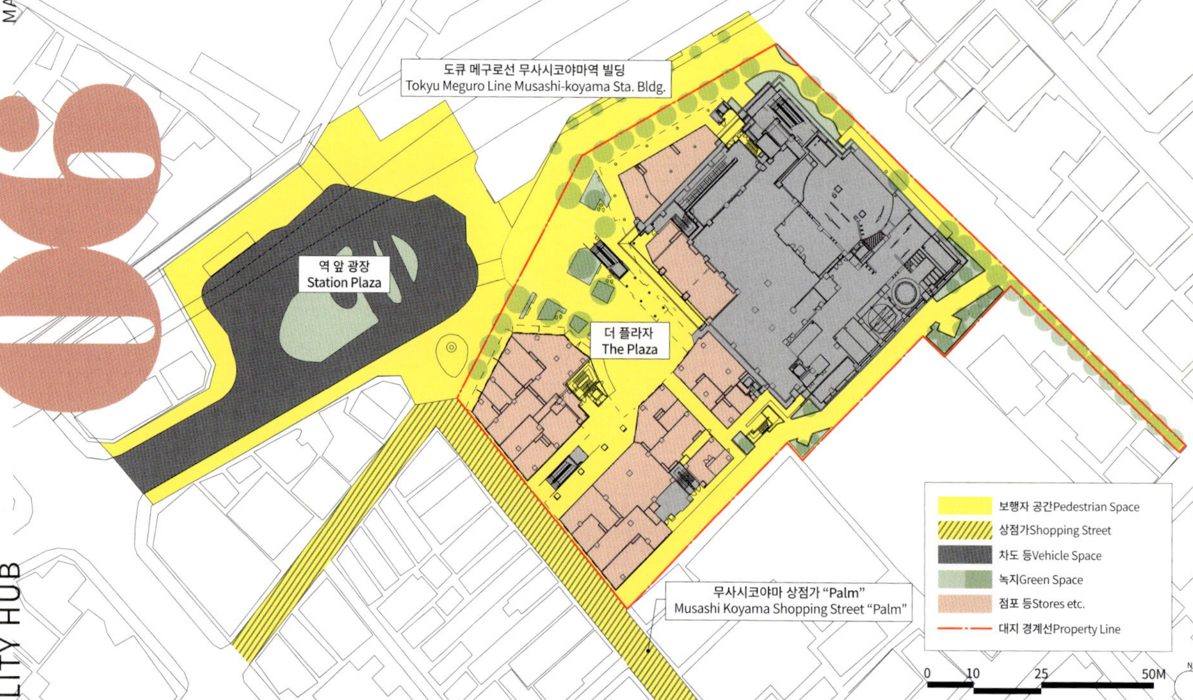

1층 평면도 및 배치도　1st floor plan and site plan

더 플라자 남서쪽에 인접한 무사시코야마 상점가 「파룸」. 정비 전의 모습
Musashi Koyama Shopping Street "Palm" adjacent to the southwest side of the Plaza. View before the development.

정비 후의 상점가
View of the shopping street after the development.

석양이 질 때 더 플라자를 내려다본 관경
Looking down at the Plaza at dusk.

Subway concourse created through public-private cooperation

—— **TOKYO MIDTOWN HIBIYA**   **Hibiya Arcade**

민관(官民)일체가 만드는 지하철 콘코스

도쿄 미드타운 히비야 히비야 아케이드(2018)

히비야 아케이드는 구가 소유하는 공유지와 민유지의 지하를 활용해 정비된 지하광장이다. 지구계획정리사업에 의해 생겨난 약 1200㎡의 지하광장은 지하철 치요다선 히비야역과 히비야선 히비야역의 두 역을 연결하는 배리어 프리 환승 콘코스로 활용되고 있으며, 민간개발에 의한 지하철역의 기능확충이 실현되고 있다. 또 광장을 사이에 두고 북쪽에는 공유지에 점포 등이 배치되어 있고 남쪽에는 민간부지 내에 점포가 배치되어 있어, 민관 쌍방의 활기를 띠는 단순한 지하 환승 공간을 넘어선 새로운 형태의 퍼블릭 스페이스로 형성되었다.

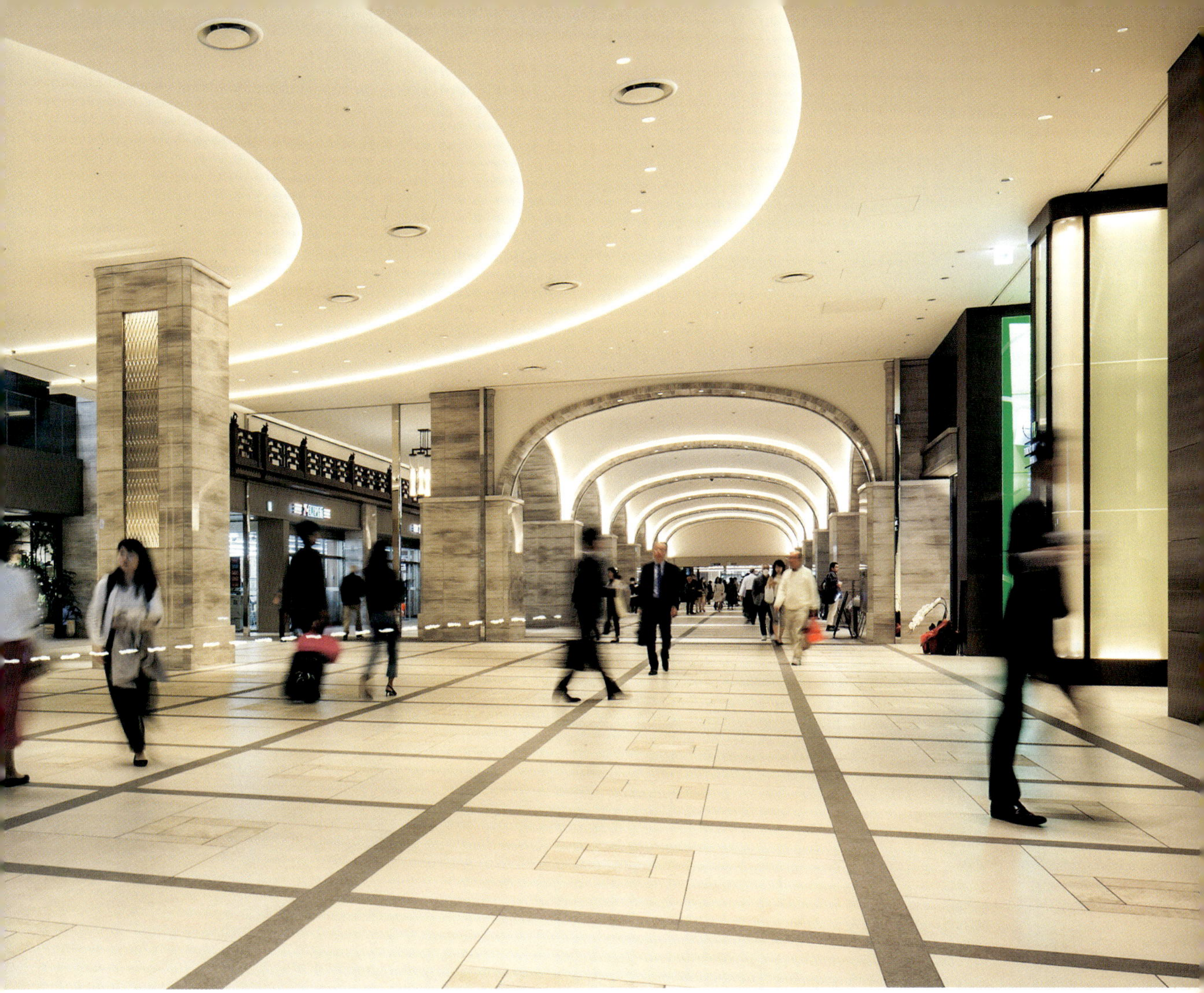

히비야 아케이드. 공유지 및 민간부지의 지하를 활용해 정비되었다.
Hibiya Arcade was developed utilizing the underground of municipal and private properties.

개발 전 배치
Site, before the development

개발 후 배치
Site, after the development

지하 1층 평면 1st basement floor plan

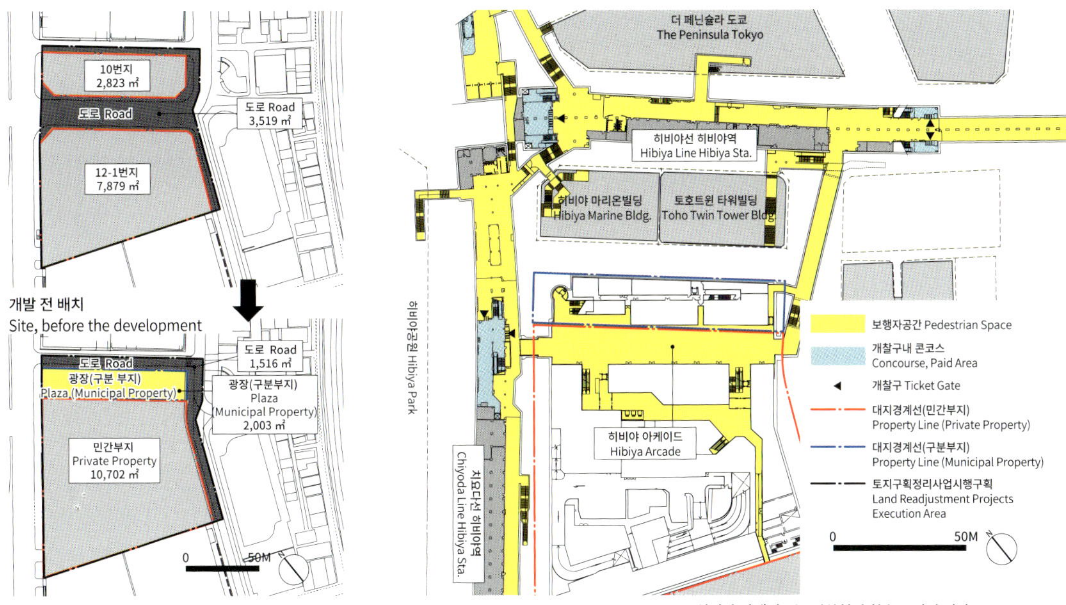

히비야 아케이드는 지하철의 환승 동선이 된다.
Hibiya Arcade acts the circulation path for subway transfer.

도쿄메트로 토자이선
Tokyo Metro Tozai Line

니혼바시 지하 단면도 Nihonbashi underground walkway section

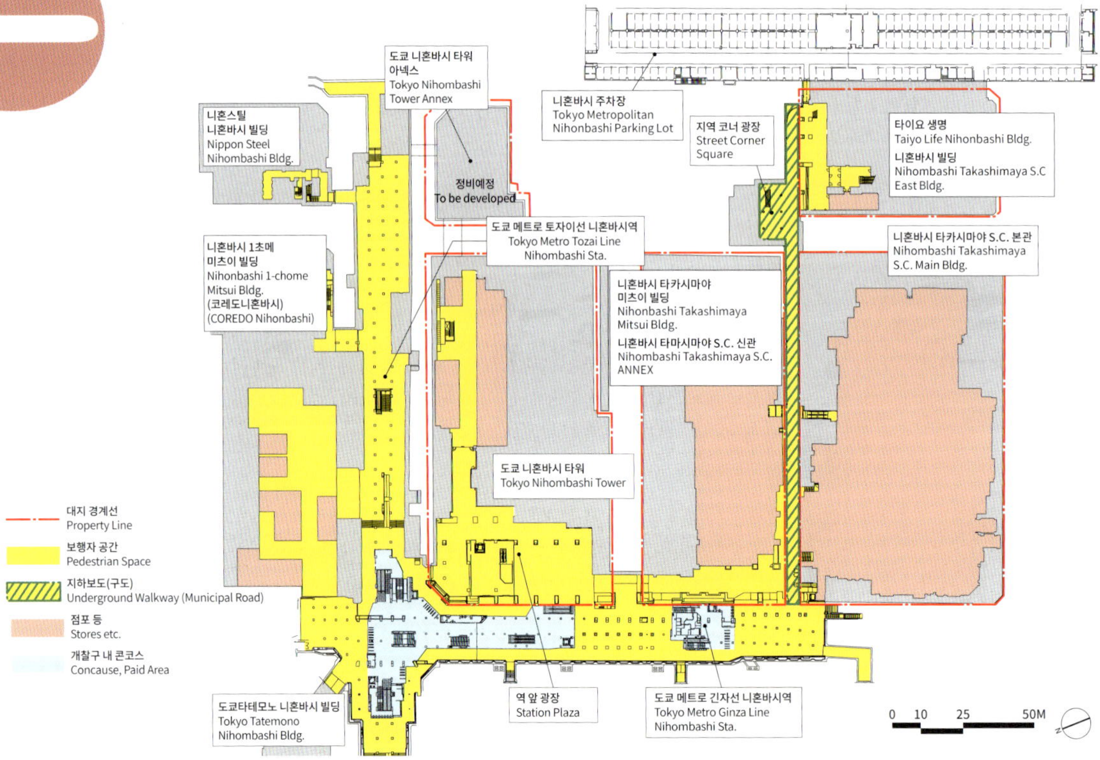

지하1층 평면   1st basement floor plan

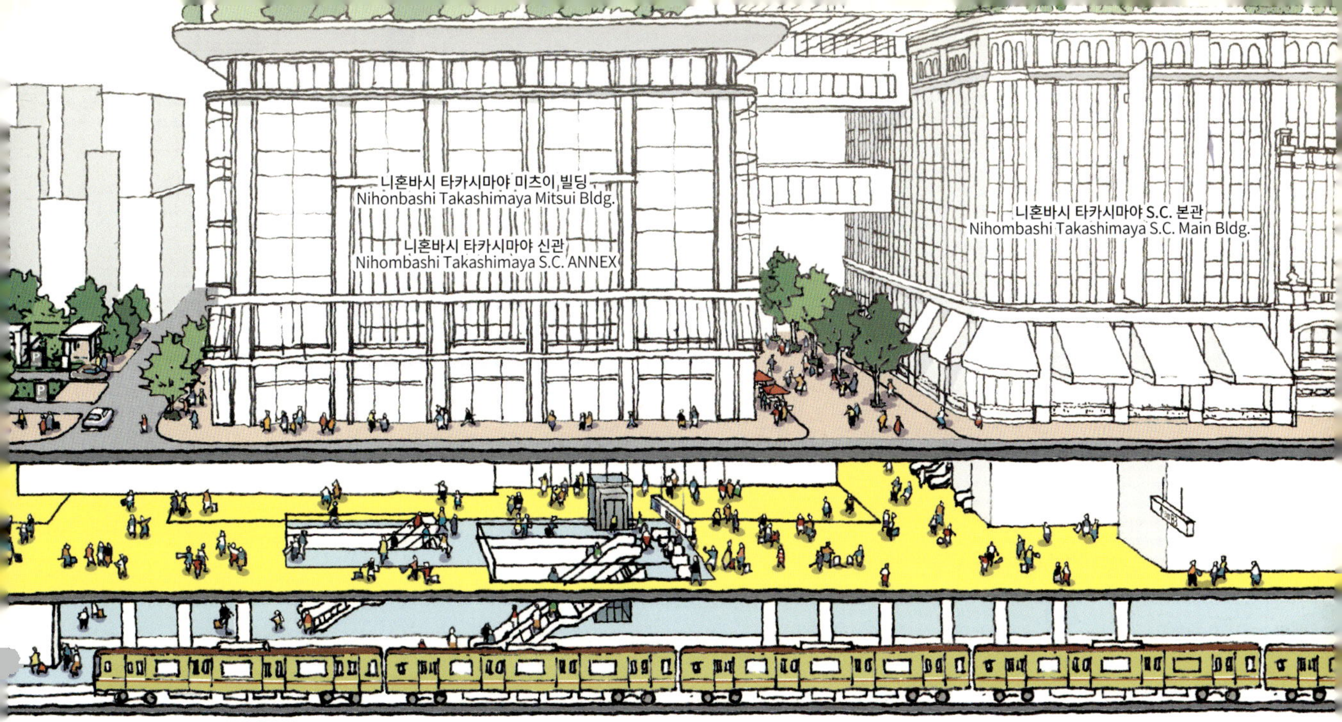

Integration of a subway station and town using private properties

—— **Subway Nihombashi Station　Underground Walkway**

민간부지를 활용한 지하철역과 도시의 일체화

지하철 니혼바시역 지하보도(2018)

도쿄 니혼바시타워(2015), 니혼바시 다카시마야 미츠이빌딩(2018), 타이요세이메이 니혼바시빌딩(2018), 니혼바시 다카시마야 S.C.(2018)

지하철 니혼바시역은 도로관리자, 철도관리자, 행정 및 다수의 개발자들 간의 조정 및 연계를 통해 공간적 제약이 많은 지하철역에서 역과 도시가 일체화되는 도시공간을 실현하고 있다. 구체적으로는 개찰구 내의 환승 동선 공간 확대나 역개찰기 증설 등으로 역 내 혼잡을 개선하고, 지상/지하를 연결하는 역광장과 배리어프리 동선의 확보 등을 통해 역과 도시를 연결하기 위한 개선이 민관의 경계를 넘어 진행되었다. 지하의 역 앞 광장은 지하철역에서는 공간확보가 어려운 재해 시 피난공간으로써 기능하는 등 방재대응력 향상에도 기여하고 있다.

좌 : 니혼바시역 주변의 지하공간. 오른쪽에는 니혼바시 다카시마야 S.C 신관의 지하입구가 보인다.
우 : 도쿄 니혼바시타워 부지 내의 역 앞 광장에서 니혼바시역 개찰구를 바라본 관경
Left: Underground space around Nihombashi Station. To the right is the underground-level entrance to Nihombashi Takashimaya S.C. ANNEX. Right: View of the Nihombashi Station ticket gate seen from the station plaza located in Tokyo Nihombashi Tower site.

Large space in front of subway station realized through public-private partnership

— **Subway Mitsukoshimae Station   Nihonbashi Underground Walkway**

민관 연계로 실현된 지하철역 앞의 대공간

지하철 미츠코시마에역 니혼바시 지하보도(2002〜)

지하보도(중앙구도 区道)(2014〜), 니혼바시 미츠이타워(2005), 무로마치 히가시 미츠이빌딩(2010),
니혼바시 무로마치 노무라빌딩(2010), 무로마치 치바긴 미츠이빌딩(2014),
무로마치 후루카와 미츠이빌딩(2014), 니혼바시 무로마치 미츠이타워(2019)

도쿄 메트로 미츠코시마에역에서 JR소부선 신니혼바시역에 걸쳐 약 500m 구간에는 지하도 정비를 통해서 지하철역과 다수의 건물을 연결하는 대공간이 만들어져 있다. 이 지하보도의 정비는 갱신 시기에 근접한 길가 건물의 재건축 기회를 이용하여 민관이 연계하여 도로의 지하공간과 길가 건물을 일체화시켜 정비한 결과이다. 도로중앙부에 있는 기존 시설 지하 콘코스 양쪽에 지하보도를 신설하고, 도로 폭 전체를 활용하여 지하공간을 확보하는 것으로 건물 지하가 지하보도에 연접하게 되어 다양한 점포들이 늘어선 쾌적한 보도공간으로 재구축되었다. 지금도 이러한 지하공간 활성화의 시도들이 진행되고 있으며, 민관 연계를 통한 양질의 도시공간 만들기가 단계적으로 확대되고 있다.

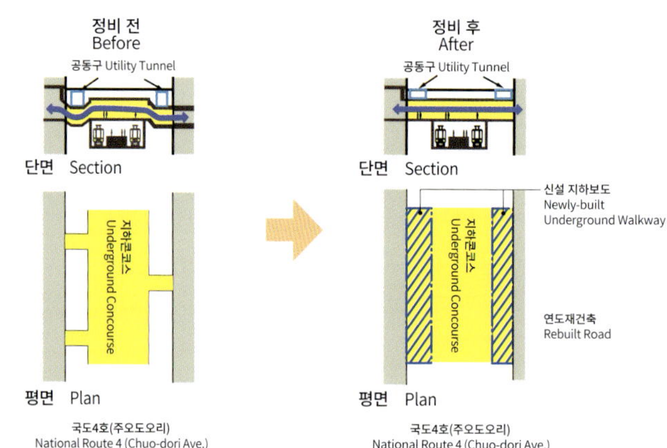

지하 콘코스와 민관 경계 사이의 공간에 지하보도를 정비하는 것으로 여유 있는 보행자 공간이 형성되고, 건물 쪽에서도 지하보도에 접한 대규모의 개구부 설치가 가능하게 된다.
The underground walkway placed in between the underground concourse and the public-private boundary allows for a wide pedestrian space, as well as a large opening facing the walkway even on the building side.

왼쪽 : 신설된 지하보도 공간. 왼쪽에는 니혼바시 무로마치 미츠이타워의 지하입구, 오른쪽에는 기존의 지하 콘코스가 보인다.
오른쪽 : 무로마치 미츠이빌딩과 무로마치 치바긴 미츠이빌딩 사이의 지하보도
Left: The newly-built underground walkway. To the left is the underground-level entrance to Nihonbashi Muromachi Mitsui Tower. To the right is the existing underground concourse.
Right: Underground walkway between Muromachi Higashi Mitsui Bldg. and Muromachi Chibagin Mitsui Bldg.

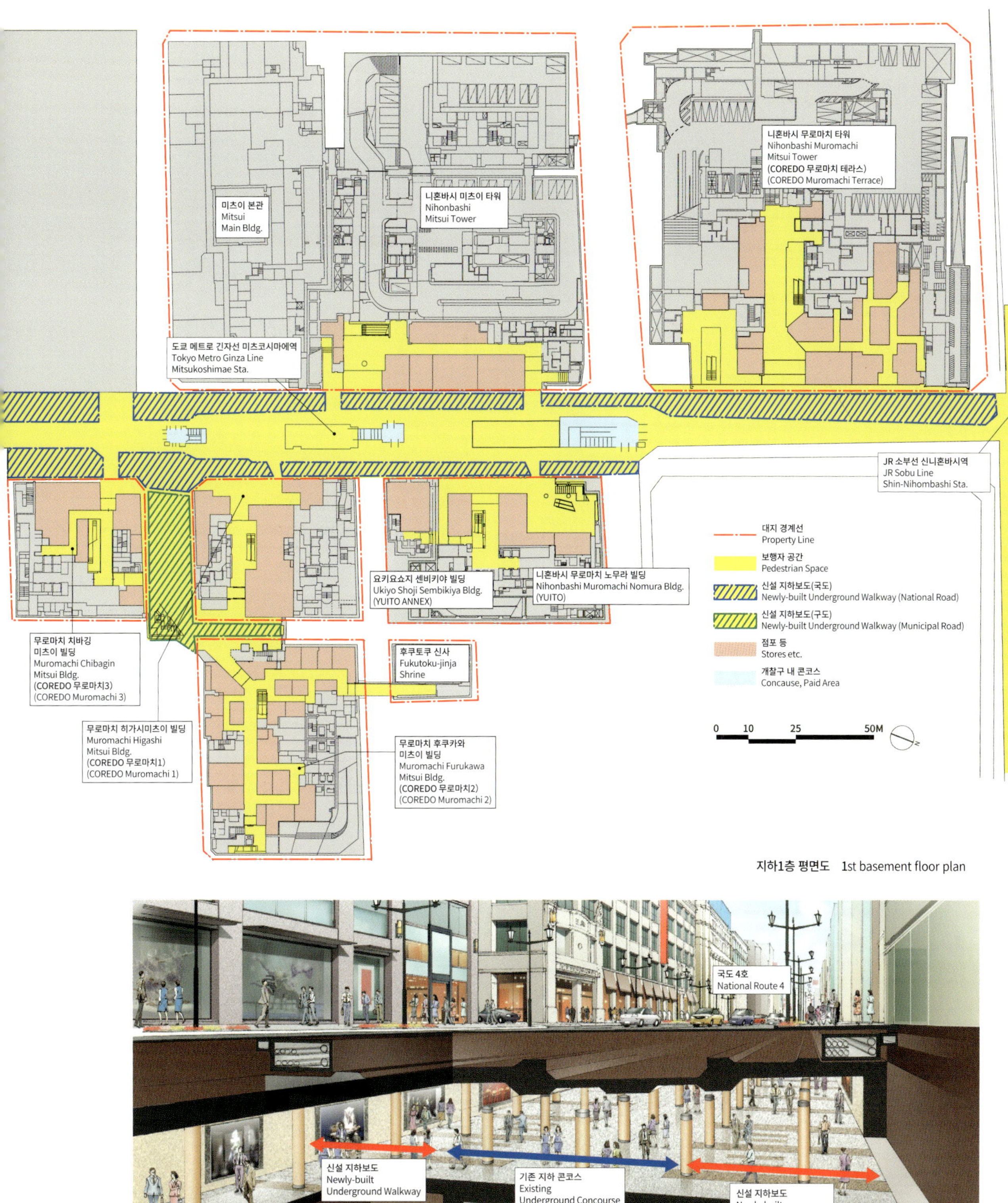

지하1층 평면도　1st basement floor plan

지하철 미츠코시마에역 니혼바시 지하보도의 단면구성
Section composition of the Nihonbashi underpass, Tokyo Metro Mitsukoshimae Station.

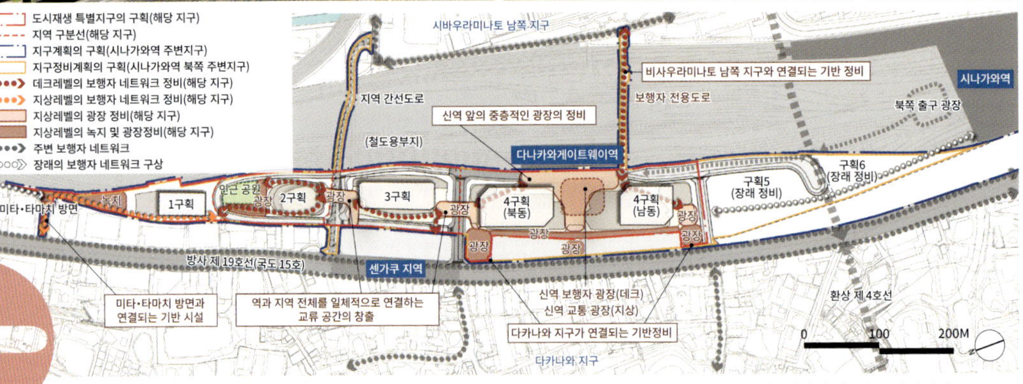

개발 후 배치　Site, after the development

왼쪽 위 : 남동쪽에서 바라본 다카나와게이트웨이역 앞의 새로운 도시 이미지
오른쪽 위: 새로운 보행자 광장의 정비 이미지
Above left: New town image of completed Takanawa Gateway Station front development project, seen from the southeast. Above right: Image of the new station pedestrian plaza.

New station-front plaza as the genesis of a pedestrian network

—— Takanawa Gateway Station　Pedestrian Plaza

ANTICIPATED PROJECT

보행자 네트워크의 기점이 되는 새로운 역 앞 광장

다카나와게이트웨이역 보행자 광장

JR야마노테선의 새로운 역 「다카나와게이트웨이역」을 중심으로 차량기지 부지를 활용한 대규모 개발로 인해 2024년에 1~4블록의 새로운 도심이 탄생할 예정이다. 남북 방향으로 약 1.6km에 달하는 새 도심을 연결하는 골격으로, 데크를 중심으로 하는 보행자 네트워크가 계획되어 있다. 이 보행자 네트워크의 중앙부에 위치한 새로운 역 앞에 정비되는 약 6500㎡의 넓은 보행자 광장은 국제교류의 장으로써 정보 발신이나 이벤트 스페이스 등으로 활용이 가능하다. 더욱이 블록 간 도로 상부에도 광장이 정비될 예정이므로, 도시 전체에 다양한 교류와 연속된 활력이 기대된다. 또한 주변지구(다카나와지구, 시바우라 미나토지구)와 연속된 보행자 네트워크 형성도 함께 검토되어 역과 도시의 광장이 일체화된 쾌적한 보행공간이 실현될 것이다.

Creating a new reginal transportation node at Tokyo Station front

Tokyo Station Regional Bus Terminal

도쿄역 앞의 새로운 지역교통결절점 만들기

도쿄역 앞 블록 버스터미널

야에스 1초메 6지구, 야에스 2초메 1지구, 야에스 2초메 나카지구

도쿄역 야에스 출구에 접한 3지구에는 도쿄역 앞 지구 도시 만들기 가이드라인에 의거하여 야에스 지하도를 통한 도쿄역과 지하철 긴자선 교바시역을 연결하는 보행자 네트워크를 정비하는 도시계획이 결정되어 있다. 보행자 네트워크 정비는 지하뿐만 아니라 지상 수준에서도 여러 블록으로 연결하는 광장공간과 보행자 네트워크가 계획되어, 공공공간과 민간개발이 연계하여 원활한 보행자 네트워크가 완성될 예정이다. 또 도쿄역 야에스의 역 앞 광장 기능을 보완하는 역할로써 세 블록을 연계해 개발 계획지 내에 버스터미널을 정비하는 것도 도시계획으로 결정되어서, 한층 도쿄역 앞의 교통결절 기능이 강화될 것이다.

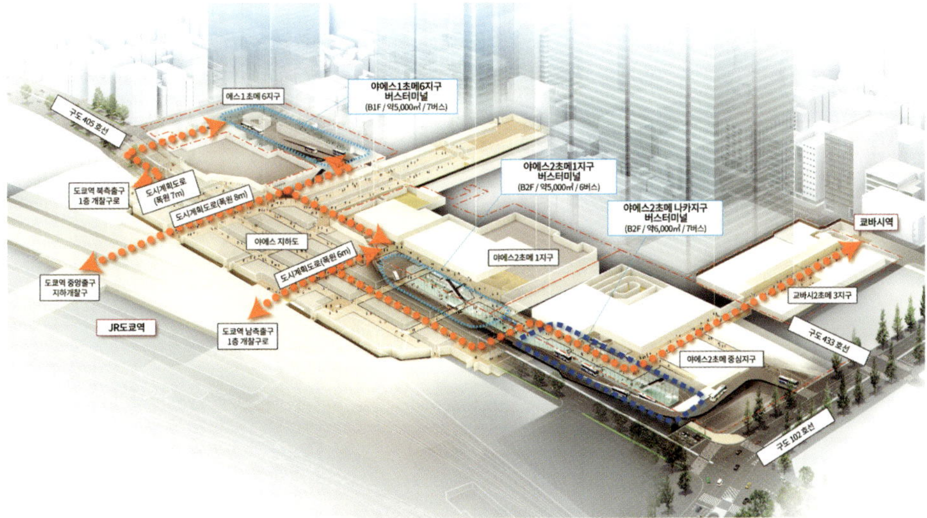

도쿄역 앞 지구 버스터미널의 지하 보행자 네트워크 정비 이미지
Image of underground pedestrian network development at Tokyo Station Regional Bus Terminal.

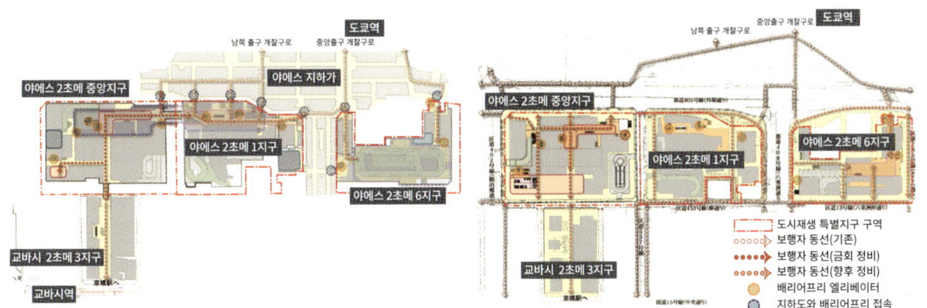

개발 전의 지하평면 Basement floor plan, after the development

개발 후의 지하평면 Ground floor plan, after the development

BARGAIN

SEAMLESS LINKAGE

A series of spaces that
form a continuous network

원활한 네트워크를 형성하는
퍼블릭 스페이스

Deck woven in phases through multiple city blocks
—— Area around Osaki Station　Deck Network

여러 블록이 단계적으로 연결된 데크

오사키역 주변지구 데크 네트워크(1987 ～)

오사키역 주변지구는 메구로강 유역의 대규모 공장용지의 토지이용 전환을 계기로 해서 단계적으로 개발되어, 복수의 블록을 관통해 역에서 방사선으로 연결되는 광범위한 데크 네트워크가 형성되었다. 부도심 개발이 착수된 당초에는 역 서쪽의 Think Park, 동쪽의 오사키 뉴시티, 게이트 시티 오사키까지의 계획이 었지만, 데크 네트워크는 지속적으로 개발 때마다 연결되어 주변지역에까지 연결되었다. 결과적으로 보행자는 차량 동선과 교차를 피해 역에서 주변지역으로 자연스레 액세스가 가능하게 되었다. 공공용지 내에는 구도(区道)로써, 민지(民地) 내에는 도시계획상의 도시시설로써 연결되는 것으로, 넓은 폭의 풍부한 보행자 공간을 실현하고 있다.

Think Park앞의 데크는 민관 경계를 넘어 일체적인 이용이 가능하도록 정비되었다.
The deck in front of ThinkPark was developed for an integrated use beyond the public-private boundary.

아트 빌리지
오사키 뷰 타워
Art Village Osaki
View Tower
2007
오사키 아트 빌리지
오사키 센트럴 타워
Art Village Osaki
Central Tower
2007
오사키 프론트 타워
Osaki Front Tower
2005
오사키 센터 빌딩
Osaki Center Bldg.
2009
오사키 뉴 시티
Ohsaki New City
1987
오사키 뉴 시티
JR Osaki Sta.
게이트 시티 오사키
(웨스트 타워)
Gate City Ohsaki
(West Tower)
1998
메구로강 Meguro River
일본 변리사회관
Nihon Zeirishi Kaikan
게이트 시티 오사키
(이스트 타워)
Gate City Ohsaki
(East Tower)
ThinkPark
2007
다이와 로이넷
호텔 도쿄 오사키
Daiwa Roynet Hotel Tokyo-Osaki
2007
NBF 오사키 빌딩
NBF Osaki Bldg.
2011
오사키 위즈 시티
Osaki Wiz City
2014
오사키 웨스트 시티 타워즈
Osaki West City Towers
2009
보행자 공간(데크 및 지상 레벨)
Pedestrian Space (Deck and Ground Level)
개찰구 내 콘코스
Concouse, Paid Area
1998 준공연도 Completion Year
블록 경계·대지 경계선
City Block Boundary / Site Boundary etc.
배치　Site　오사키역 주변지구의 보행자 네트워크
Pedestrian network around Osaki Station.
0　40　100　200M
N

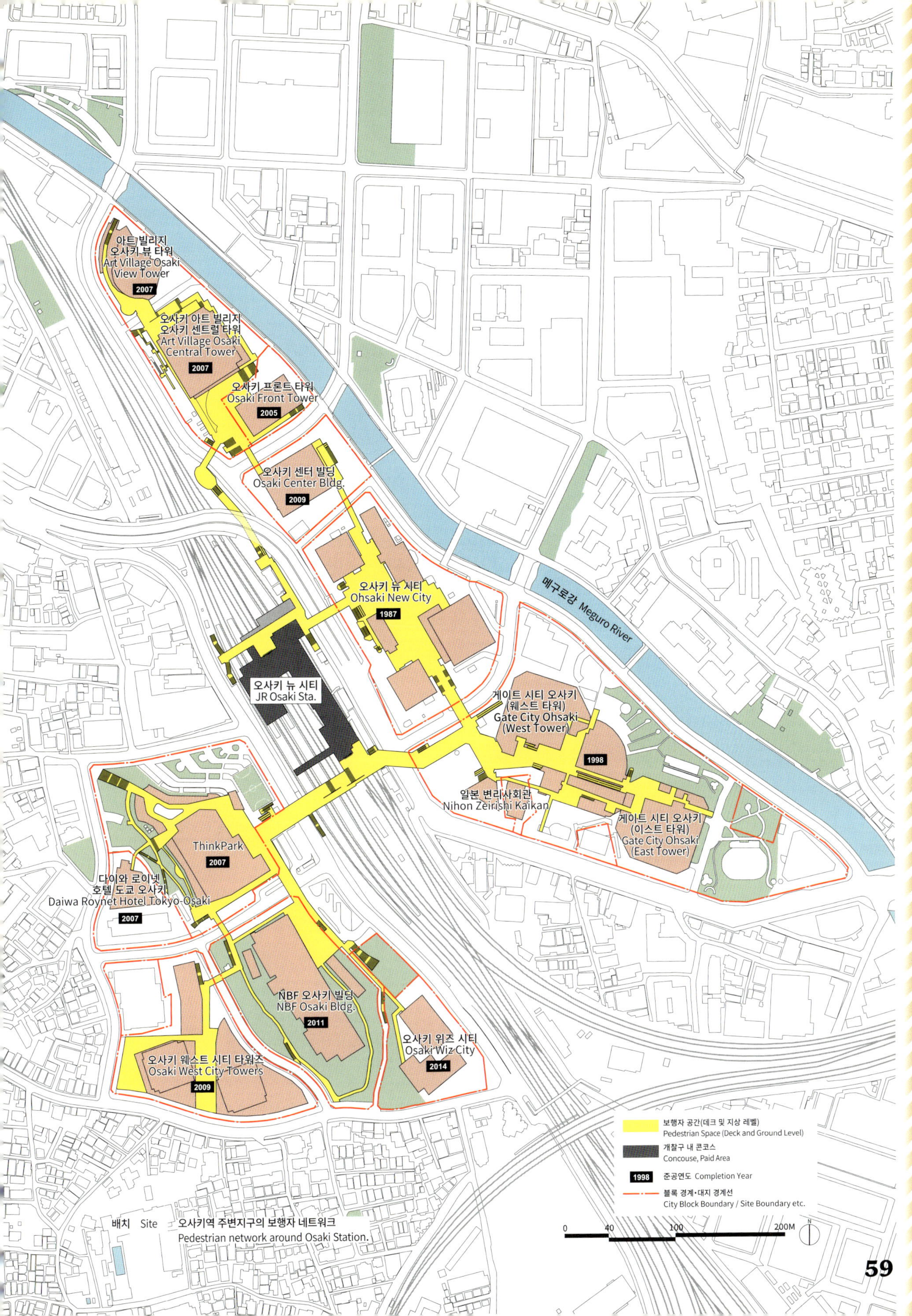

59

Multi-layered corridor connecting the valley

Shibuya Station Central District　Pedestrian Network

분지를 연결하는 입체 회랑

시부야역 중심지구 보행자 네트워크

시부야역 주변의 보행자 네트워크는 분지의 약점을 장점으로 만든 퍼블릭 스페이스다. 민지 내의 보행자 동선, 도로 점용 데크, 횡단육교가 수직 동선인 Urban Core를 매개로 해서 주변지역으로 뻗어나가, 국도 246호선과 메이지 도로 등의 주 간선도로를 넘어 역과 도시를 연결하는 광범위의 데크 네트워크가 만들어졌다. 분지의 바닥에 위치한 역의 동서쪽을 상부에 연결해, 도시의 순환체계를 크게 변화시킬 입체 회랑의 완성을 눈앞에 두고 있다. 현재 정비 중인 도쿄메트로 긴자선 상부[스카이 데크]가 완성되면 보다 수평적으로 역의 동서쪽이 연결된다.

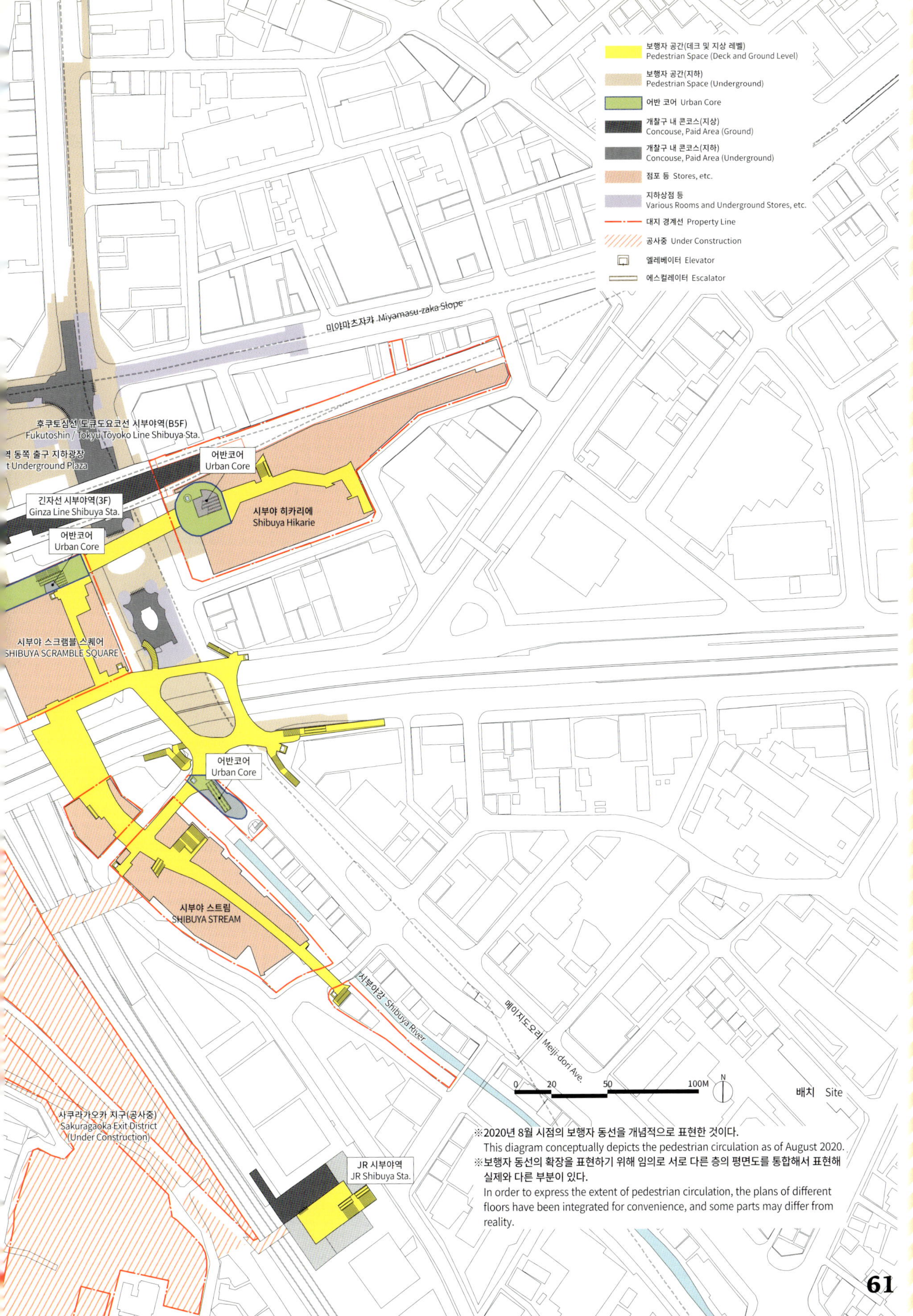

보행자 공간(데크 및 지상 레벨)
Pedestrian Space (Deck and Ground Level)
보행자 공간(지하)
Pedestrian Space (Underground)
어반 코어 Urban Core
개찰구 내 콘코스(지상)
Concouse, Paid Area (Ground)
개찰구 내 콘코스(지하)
Concouse, Paid Area (Underground)
점포 등 Stores, etc.
지하상점 등
Various Rooms and Underground Stores, etc.
대지 경계선 Property Line
공사중 Under Construction
엘레베이터 Elevator
에스컬레이터 Escalator
미야마츠자카 Miyamasu-zaka Slope
후쿠토신선 / 도큐도요코선 시부야역(B5F)
Fukutoshin / Tokyu Tōyoko Line Shibuya Sta.
역 동쪽 출구 지하광장
t Underground Plaza
어반코어
Urban Core
긴자선 시부야역(3F)
Ginza Line Shibuya Sta.
어반코어
Urban Core
시부야 히카리에
Shibuya Hikarie
시부야 스크램블 스퀘어
SHIBUYA SCRAMBLE SQUARE
어반코어
Urban Core
시부야 스트림
SHIBUYA STREAM
시부야강 Shibuya River
메이지도오리 Meiji-dori Ave.
사쿠라가오카 지구(공사중)
Sakuragaoka Exit District
(Under Construction)
JR 시부야역
JR Shibuya Sta.
0   20   50   100M
N
배치   Site

※2020년 8월 시점의 보행자 동선을 개념적으로 표현한 것이다.
This diagram conceptually depicts the pedestrian circulation as of August 2020.
※보행자 동선의 확장을 표현하기 위해 임의로 서로 다른 층의 평면도를 통합해서 표현해
실제와 다른 부분이 있다.
In order to express the extent of pedestrian circulation, the plans of different
floors have been integrated for convenience, and some parts may differ from
reality.

위 : 시부야 마크시티에서 본 진구도오리 상공의 통로
아래 : 시부야 스트림과 국도 246호 상공의 통로 접속부
Above: Walkway above the Jingu-dori Ave. seen from SHIBUYA MARK CITY.
Below: The connection between SHIBUYA STREAM and a walkway above National Route 246.

Urban Core that facilitates vertical movement in the network

네트워크에 입체적인 회유성을 만드는 Urban Core

시부야에는 입체 회랑을 지지하고 있는 Urban Core라고 불리는 대규모의 수직 동선 공간이 있다. 이는 지하 역, 지상의 버스터미널, 고가 역의 이용자를 흡수하여 수평적으로 연결하는 역할을 하고 있다.

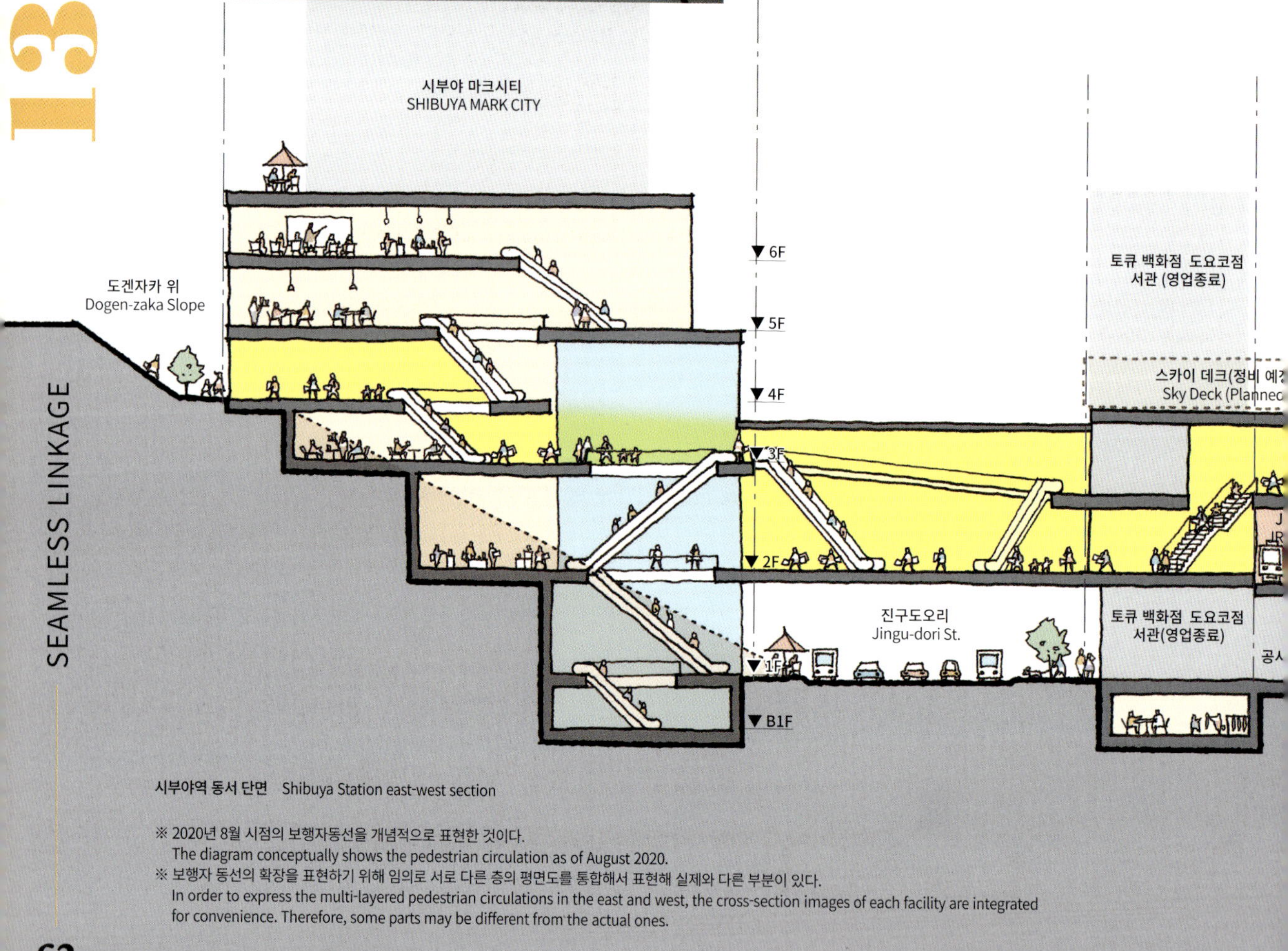

13

시부야역 동서 단면   Shibuya Station east-west section

※ 2020년 8월 시점의 보행자동선을 개념적으로 표현한 것이다.
The diagram conceptually shows the pedestrian circulation as of August 2020.

※ 보행자 동선의 확장을 표현하기 위해 임의로 서로 다른 층의 평면도를 통합해서 표현해 실제와 다른 부분이 있다.
In order to express the multi-layered pedestrian circulations in the east and west, the cross-section images of each facility are integrated for convenience. Therefore, some parts may be different from the actual ones.

메이지도오리 상부에 연결된 시부야역 개발구 동쪽 출구 2층 데크와 시부야 스크램블 스퀘어의 접속부
The connection between SHIBUYA SCRAMBLE SQUARE and the 2nd floor pedestrian deck of Shibuya Sta. East Gate above Meiji-dori Ave.

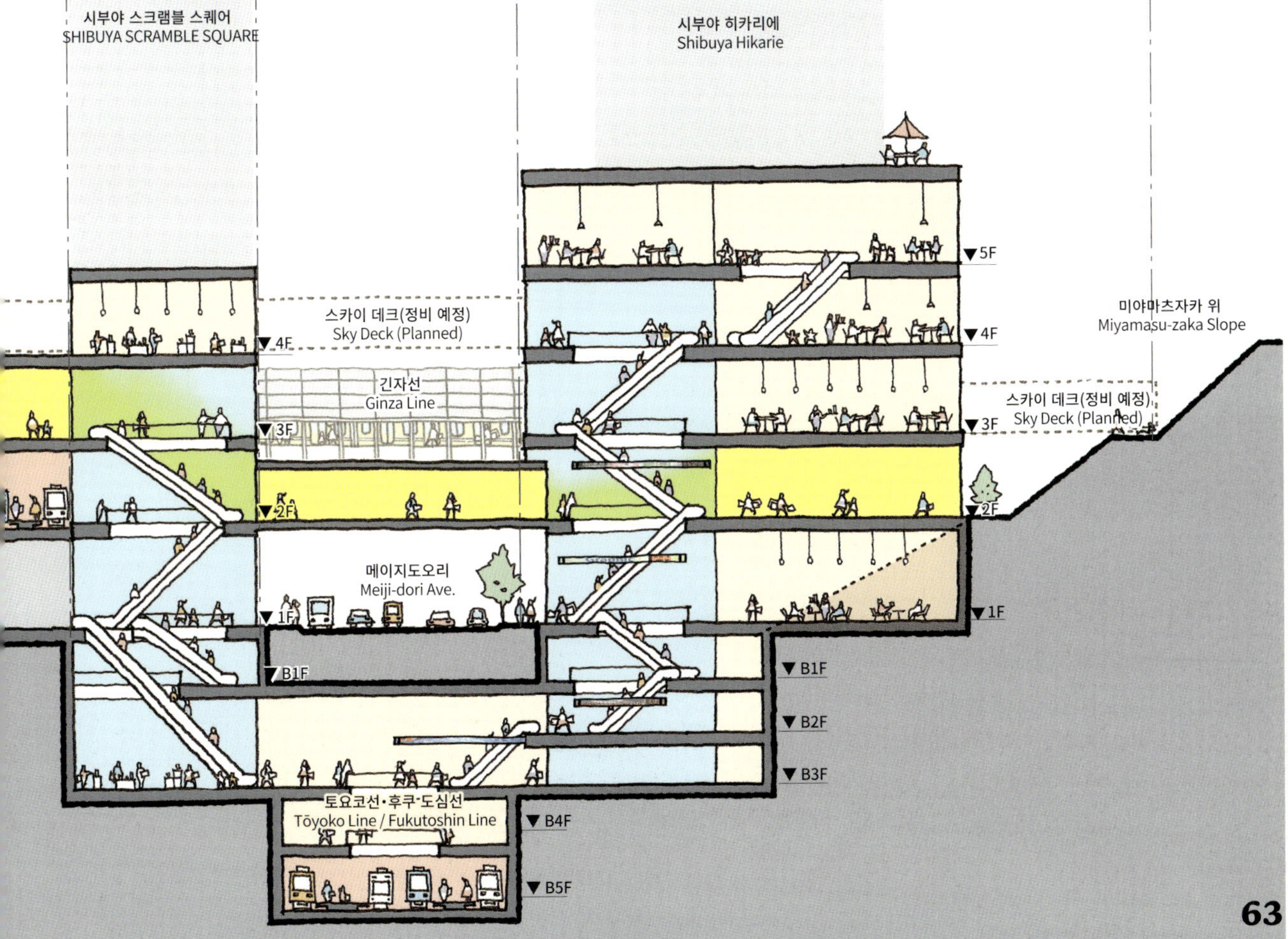

시부야 히카리에(2012)
Shibuya Hikarie

시부야 스트림(2018)
SHIBUYA STREAM

시부야 스크램블스퀘어(2019)
SHIBUYA SCRAMBLE SQUARE

시부야 후크라스(2019)
SHIBUYA FUKURAS

심리스(seamless) 네트워크의 분류

연결성 좋은 보행자 네트워크는 다양한 형태로 도심에 존재한다. 자동차나 철도와 교차하지 않도록 데크레벨로 연속하는 경우, 지상레벨의 도로를 매력적인 퍼블릭 스페이스로 만들고 있는 경우, 지하 철역에서 연결된 지하레벨의 네트워크 등, 레벨에 의해 서로 다른 형태로 존재한다. 또한 이것을 표현 하는 수법이나 관련하는 법제도 다양하다

A TYPE

데크레벨
Deck level

- 시부야역 중심지구 보행자 네트워크
- 오사키 주변 데크 네트워크

보행자는 자동차 교통과 교차되지 않도록 데크레벨로 복수 블록을 관통하는 보행자 네트워크를 형성한다.

B TYPE

지상 스트리트
Street

- 마루노우치 나카도오리
- 도쿄 미드타운 히비야 스텝 광장 등
- 아카사카/토라노몬 녹도(緑道) 구상
- 니혼바시 갤러리아

지상레벨의 도로에 대해서 복수의 블록 개발이 매력적인 퍼블릭 스페이스를 제공 하거나, 도로 자체를 보행자 전용 도로로 만드는 것으로 매력 넘치는 스트리트가 된다.

COLUMN

C TYPE

공공도로
Through passage

- 시부야 파르코 휴릭빌딩 나카시부도 오리
- 도쿄 스퀘어가든 공공통로 교바시 에도그랑 공공통로

도로를 개발의 일부로 포함해 보행자 네트워크로써의 기능을 유지하면서 개발의 중심에 위치한 공공통로로 활력 있는 퍼블릭 스페이스로 변화시키고 있다.

D TYPE

지하레벨
Underground level

- 다이마루유 지구 보행자 네트워크
- 오테마치 타워 프라자
- 오테마치 플레이스
- 시오도메 사이트 지하보도와 썬큰 광장
- 롯본기1초메역 주변 네트워크
- 토라노몬힐즈역

지하레벨에서는 지하철역과 지하보도, 그리고 지하보도와 접속된 민간개발의 지하층의 광역적인 네트워크가 펼쳐져있다.

원활한 네트워크를 형성하는
퍼블릭 스페이스

MAP p.169

SEAMLESS LINKAGE

마루노우치 나카도오리, 보행자 천국 이용시의 모습. 많은 사람들의 왕래로 거리의 활기를 만들고 있다.
Marunouchi Naka-dori Street during traffic-free use. The area is filled with people walking around.

도심의 활기를 유도하는 변화된 금융가

마루노우치 나카도오리(2002)

도쿄역 정면으로 펼쳐진 길이 약 1,200m의 스트리트. 업무기능 중심이었던 지역에 다양한 도시기능을 도입해 활력을 만든 것이 마루노우치 나카도오리이다. 비지니스 스트리트이기 때문에 야간이나 휴일에 보행자가 적었던 도로를 재정비해 쾌적한 가로를 만듦과 동시에 가로에 면한 점포 등을 유도하여 항상 사람들의 왕래가 있는 활기 있는 스트리트로 새로 태어났다.

Financial district transformed by inducing liveliness

Marunouchi Naka-dori Street

민간부지
Private Property

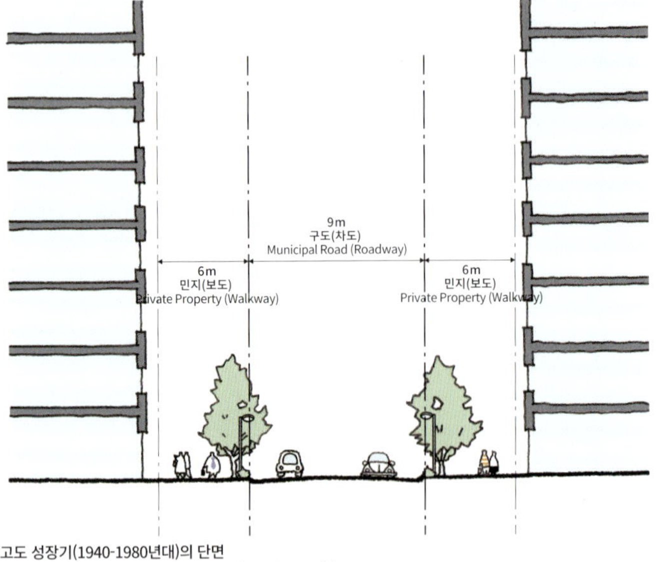

고도 성장기(1940-1980년대)의 단면
Section during rapid-growth period (1940's-1980's)

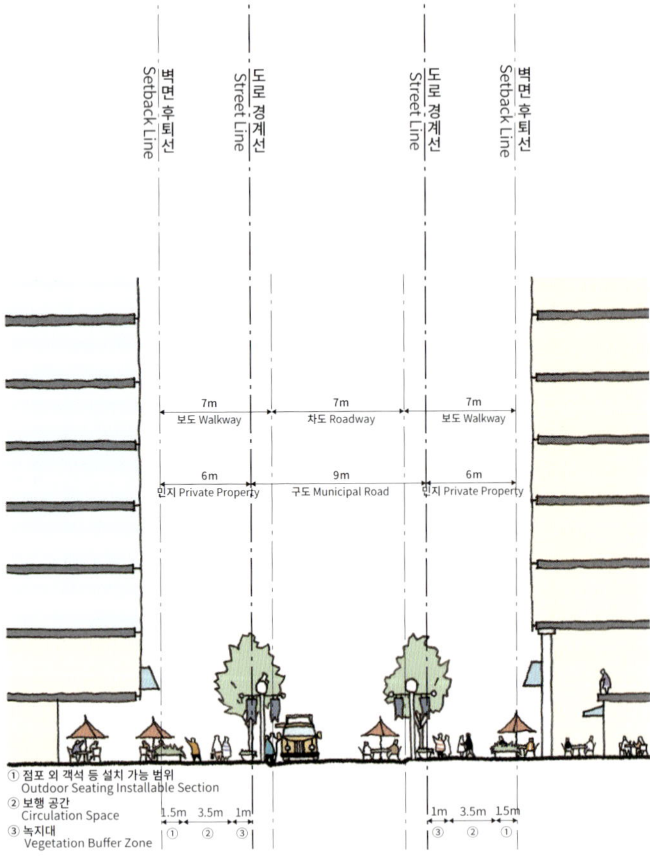

1980년대—현재의 단면
Section during 1980's to present

BEFORE

고도 성장기의 마루노우치 나카도오리
Marunouchi Naka-dori Street during the period of
rapid economic growth.

Pursuing the creation of pleasant pedestrian space

걷는 것이 즐거운 가로공간 만들기

나카도오리의 단면구성은 시대와 함께 변화해 왔다. 고도 경제성장기 때 6m(보도＝민간용지)—9m(차도＝구도(区道))—6m(보도＝민간용지)였던 도로공간은 도시만들기 가이드라인에서 「어메니티/활력축」으로 정의된 것을 계기로 7m(보도＝구도＋민간용지)—7m(차도＝구도)—7m(보도＝구도＋민간용지)로 되어 보행자 공간이 확장되었다. 더욱이 2018년 1월에 개정된 지구계획의 운용 기준에 의해, 민간용지 내의 식재 부분 1.0m와 통행용의 유효폭 3.5m를 확보하면서 남은 건물 쪽 1.5m에 대해서는 음식점의 점포객석 등으로 설치가 가능하게 되었다. 보도/차도/보도상 공지의 운용을 포함해 일체적으로 디자인하는 것으로 가로의 활력을 창출하고 있다.

AFTER

현재의 마루노우치 나카도오리.
Current Marunouchi Naka-dori Street.

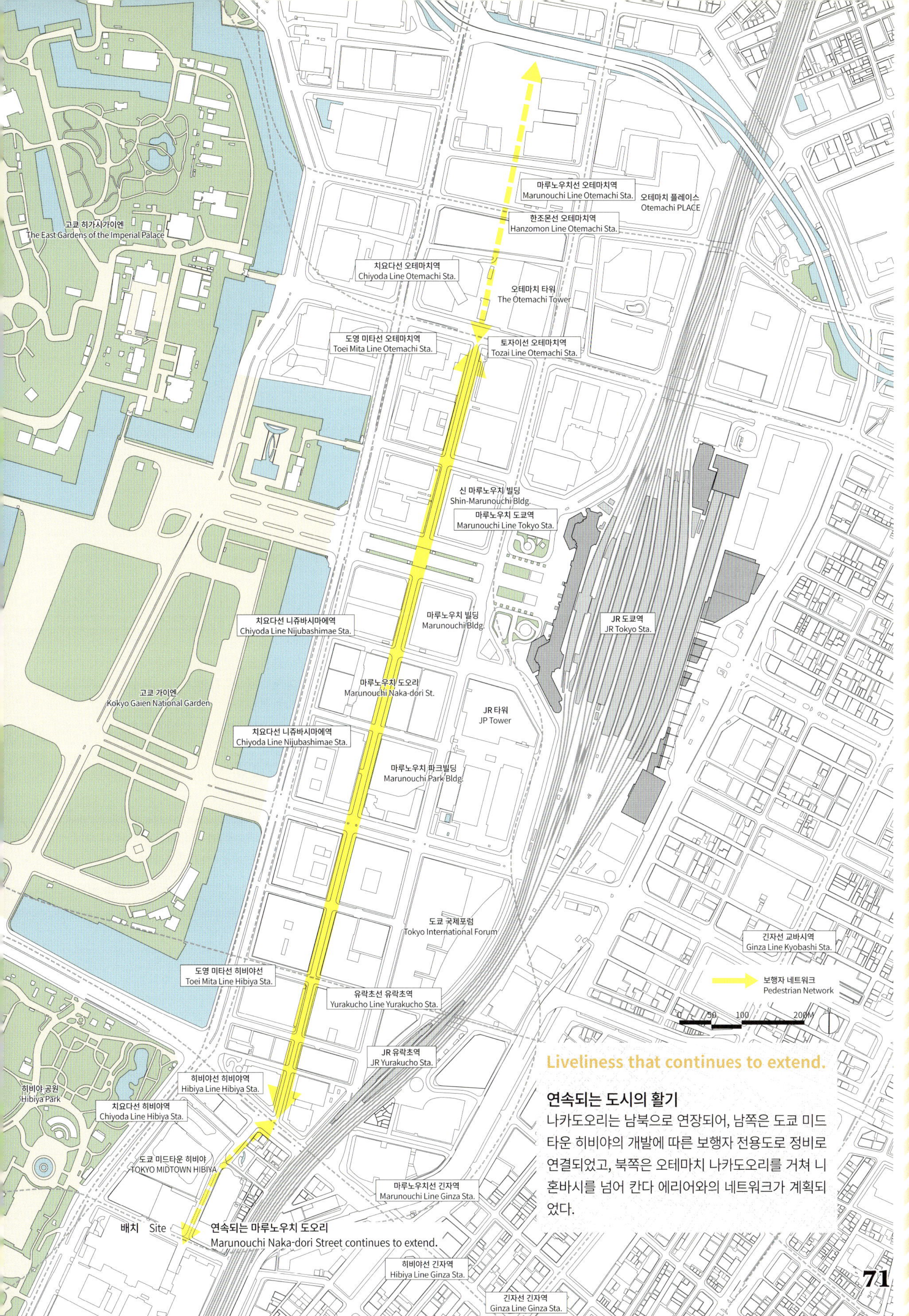

Liveliness that continues to extend.

연속되는 도시의 활기

나카도오리는 남북으로 연장되어, 남쪽은 도쿄 미드
타운 히비야의 개발에 따른 보행자 전용도로 정비로
연결되었고, 북쪽은 오테마치 나카도오리를 거쳐 니
혼바시를 넘어 칸다 에리어와의 네트워크가 계획되
었다.

15

Transforming roads into a plaza

TOKYO MIDTOWN HIBIYA Hibiya Step Square, etc.

도로를 광장으로 만들다

도쿄 미드타운 히비야 히비야 스텝광장 등(2018)

토지구획정리사업으로 정부 소유지와 민간 소유지가 일체적으로 정비되어 도로였던 공간이 계단형의 보행자 전용 광장인「히비야 스텝광장」으로 다시 태어났다. 개발과 동반해서 구도 131호 보도의 확장과 히비야 나카도오리(구도로 136호)가 보행자 전용도로가 됨으로써 동시에 기존의 광장(현재의 히비야 고질라 스퀘어)과도 일체화되어 보행자를 위한 공간이 대폭 늘어났다. 하루미 도오리까지 형성되어 있던 마루노우치 나카 도오리와도 연결되어 광역 보행자 네트워크의 일부가 되었다.

MAP p.188

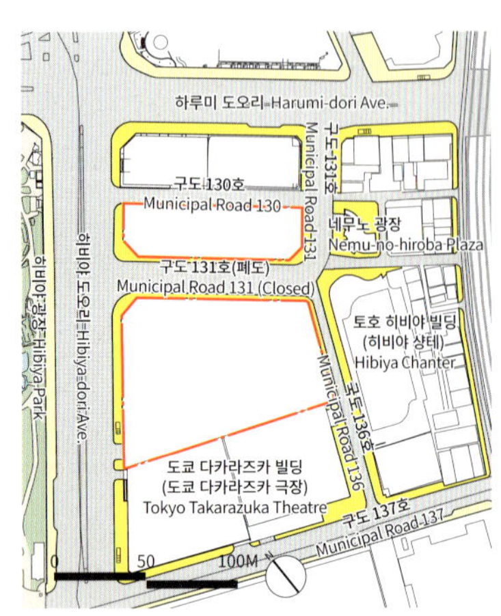

개발 전 배치 Site, before the development

개발 후 배치 Site, after the development

SEAMLESS LINKAGE

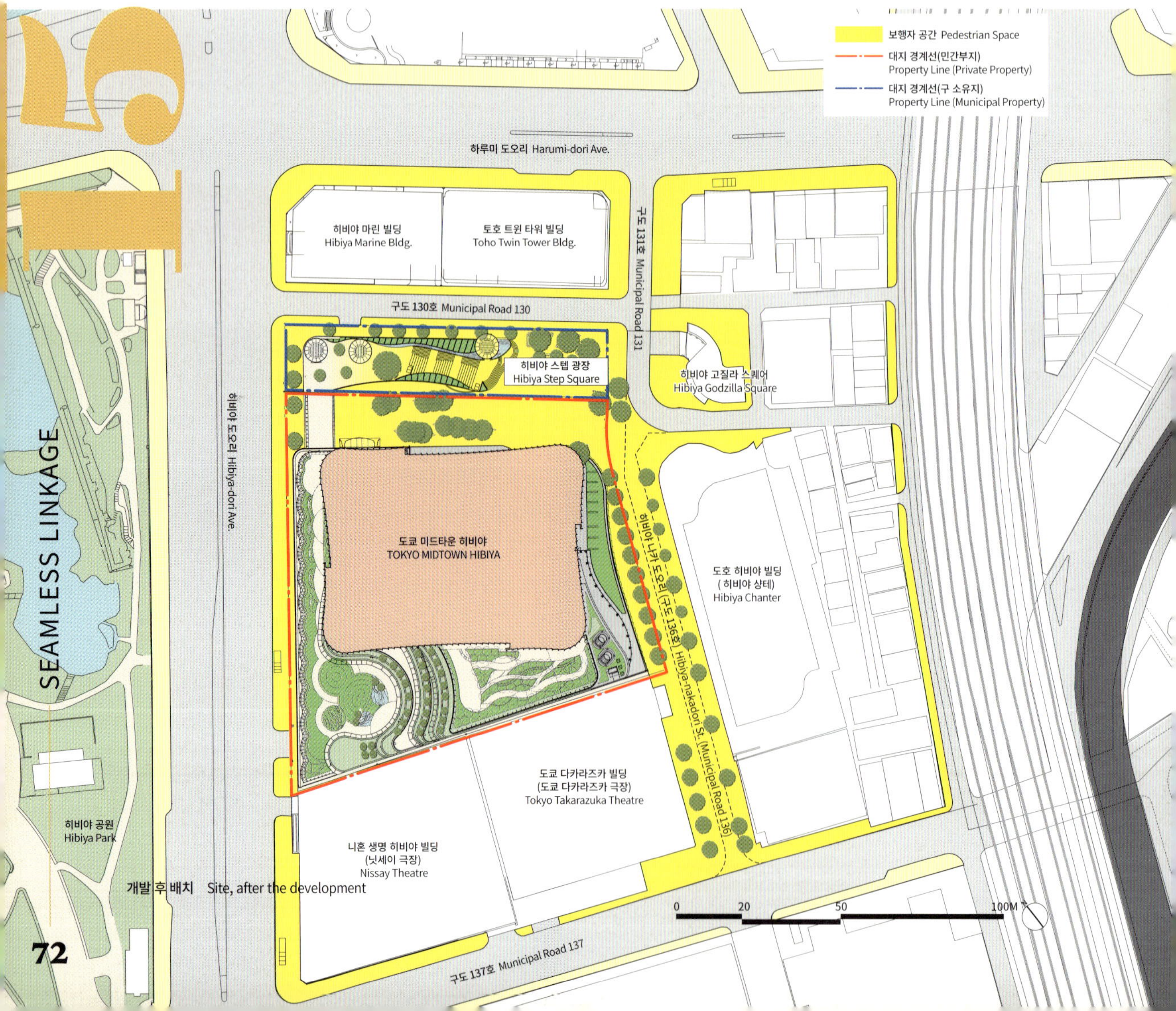

치요다구 소유지와 민간용지가 일체적으로 정비된 히비야 스텝광장
Chiyoda City property and private property were developed integrally to create Hibiya Step Square.

보행자 전용 도로가 된 나카 도오리(구도로 136호)를 바라봄
View toward Hibiya Naka-dori Street (Municipal Road 136) turned into a pedestrian-only street.

고층동에서 동쪽의 인터시티 가든과 가로수길의 점포를 바라봄
View toward Intercity Garden on the east side stores along the green path from the high-rise building.

도심에 대규모 녹도(綠道)를 만들다

아카사카/토라노몬 녹도 구상
아카사카 인터시티 AIR보도 상공지 등(2017)

세계 각국의 대사관과 외국기업이 모여 있어 국제성/다양성이 풍부한 아카사카/토라노몬에는 이 장소를 찾는 사람들을 맞이하는 쾌적한 녹지 공간이 만들어졌다. 이곳은 도심의 녹지재생 관점에서 전개한 「아카사카/토라노몬 녹도 구상」을 통한 도시만들기를 계기로 탄생하게 되었다.

Creating a grand green corridor in central Tokyo

Akasaka-Toranomon Green Corridor Project
Akasaka Intercity AIR Privately Owned Sidewalk Open Space, etc.

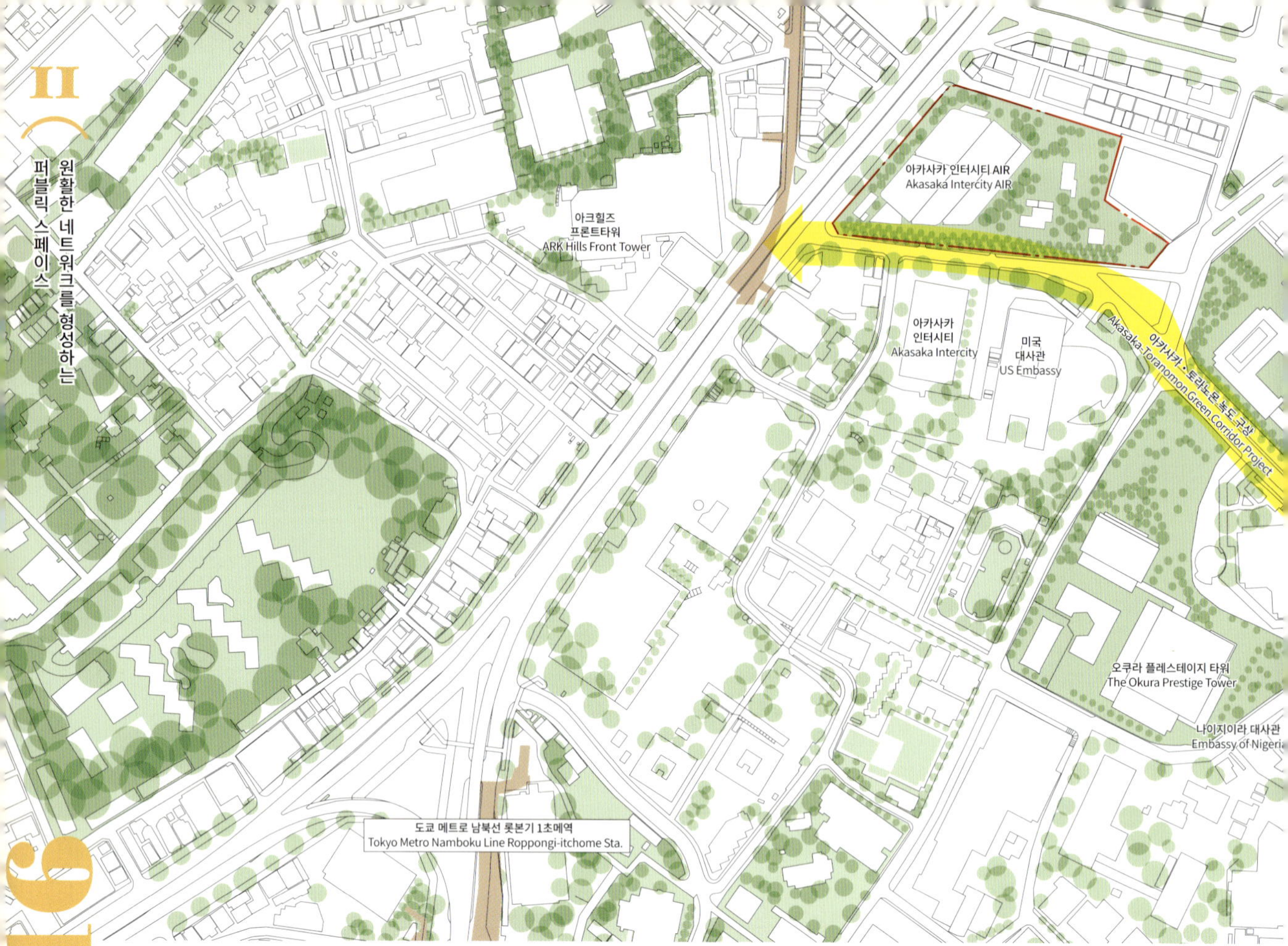

96

배치　Site　아카사카/토라노몬 녹도 구상에서는 아카사카 인터시티 AIR에서 토라노몬 힐즈까지 약 850m 구간에 풍부한 녹지의 가로공간 창출을 목표로 하고 있다.

An avenue comprising of trees on private land and large privately owned open spaces create a green hub

부지 내의 가로수 정비와 대규모 공개 공지에 의한 녹지의 거점 창출

녹도 구상을 감안해 아카사카 인터시티 AIR은 부지 내의 보도상 공지에 3열의 식재를 연장해 약 200m의 가로수길을 정비했으며, 부지 내 대규모의 공개 공지를 녹지와 합쳐서 아카사카/토라노몬 간 녹지 네트워크의 기점이 되는 풍부한 녹화 공간을 형성하였다. 대형 녹도에 걸맞은 가로수길을 형성하기 위해 도로관리상 가지치기가 필요 없는 고목 식재를 보도상 공지 내에 국한함으로써 관리하기 편리한 녹지 공간을 만들었다.

녹도 내의 보행자 공간 단면　Section of the pedestrian space within the green path

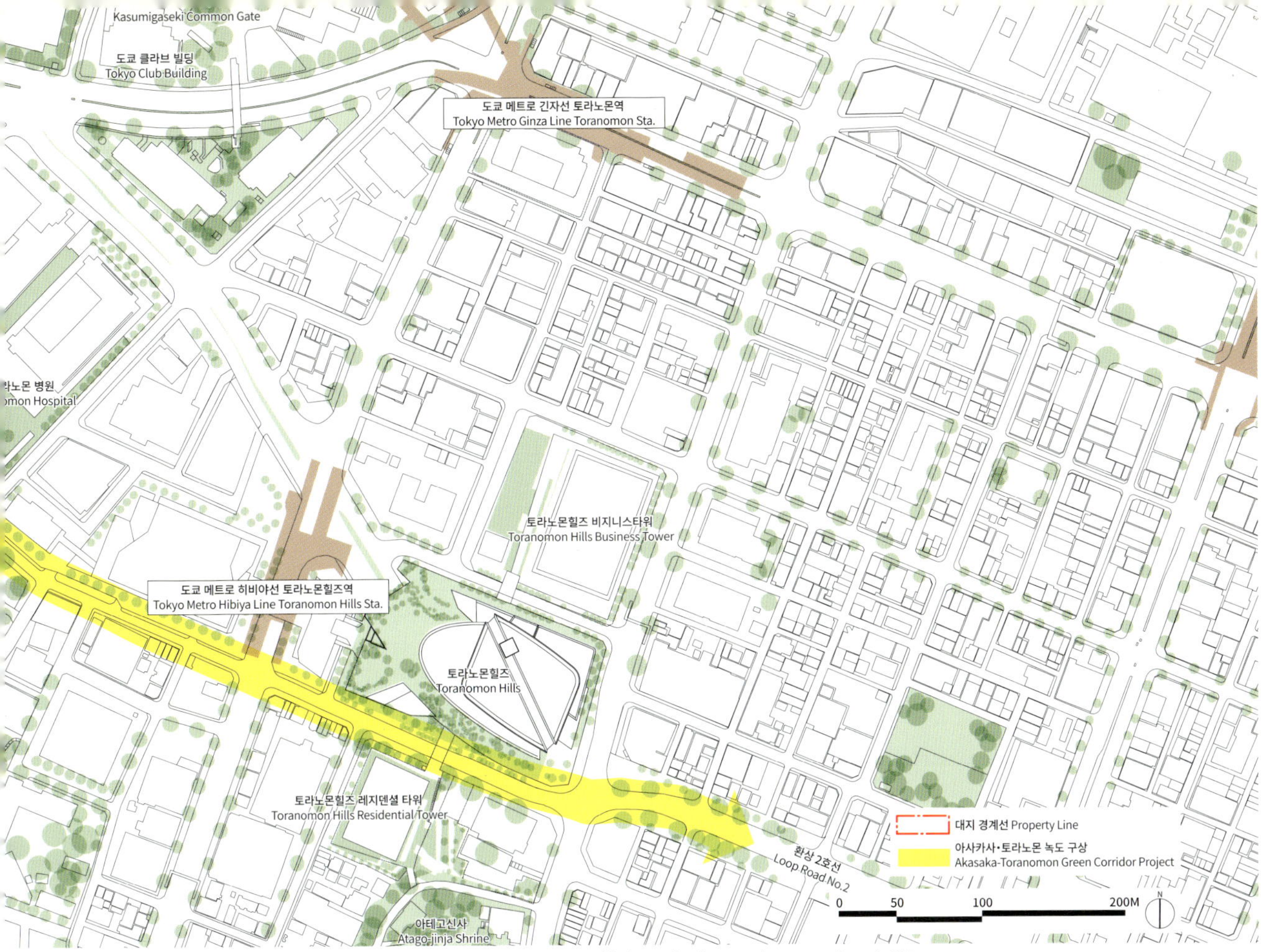

The Akasaka-Toranomon Green Corridor Project aims to create green-filled, tree-lined spaces along approximately 850m of street between Akasaka Intercity AIR and Toranomon Hills.

동쪽에서 바라봄. 녹도를 구성하는 보도상 공지에 3열로 나무를 심고, 안쪽에는 녹도를 따라 점포와 고층동을 배치했다.
View from the east. Three rows of planting in privately owned sidewalk open space form a green corridor. Toward the back stand stores along the green path and the high-rise building.

도로를 갤러리아로 만들다

니혼바시 갤러리아 (2018)

니혼바시 다카시마야 미츠이빌딩 (2018) 니혼바시 다카시마야 S.C. (2018)

니혼바시의 전통 있는 백화점으로 중요 문화재로 지정받은 「니혼바시 다카시마야 S.C 본관」의 개수와 「니혼바시 다카시마야 S.C 신관을 포함한 니혼바시 다카시마야 미츠이빌딩」의 신설을 시작으로 한 여러 가구에 걸친 개발로, 블록을 둘로 나누었던 도로가 보행자 전용도로 정비되어 마치 민용지의 갤러리아처럼 퍼블릭 스페이스로 새로 태어났다.

서쪽에서 바라봄. 왼쪽이 니혼바시 다카시마야 S.C. 신관을 포함한 니혼바시 다카시마야 미츠이빌딩, 오른쪽이 니혼바시 다카시마야 S.C 본관. 그 중간이 니혼바시 갤러리아로 정비되었다.
View from the west. On the left is Nihonbashi Takashimaya Mitsui Bldg. which includes Nihombashi Takashimaya S.C. ANNEX, and on the right is Nihombashi Takashimaya S.C. Main Bldg. The space in between was developed as Nihonbashi Galleria.

Reconfiguring the street section through integration redevelopment of vehicle circulation

차량동선의 집약과 개발을 계기로 한 도로 단면의 재정비

여러 블록의 통합개발을 통한 주차장 출입구의 통합과 도시구간 연계를 통해 수화물 반출입 공간과 차량동선을 정리하여 지하에 설치하는 등, 자동차 교통을 계획적으로 처리하여 자동차가 통행하는 도로가 사람들의 활기가 넘치는 보행자 전용 도로로 리노베이션되었다.

프로젝트를 통해 생겨난 보행자 전용도로는 유리 지붕으로 덮여 기후와 날씨에 영향을 받지 않는 갤러리아가 됨과 동시에 신구건물의 일체감 형성에도 공헌하고 있다. 가로 양쪽에는 활력을 만들어내는 점포와 지하철 출입구가 설치되어 니혼바시의 활기를 잇는 퍼블릭 스페이스가 되었다.

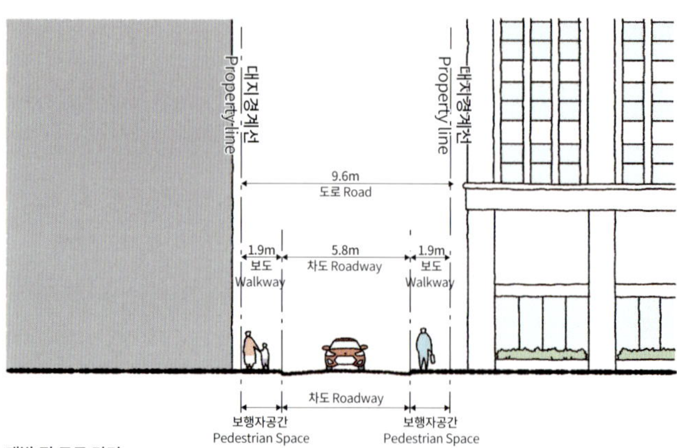

개발 전 도로 단면
Street section, before the development

개발 후 도로(니혼바시 갤러리아) 단면
Street (Nihonbashi Galleria) section, after the development

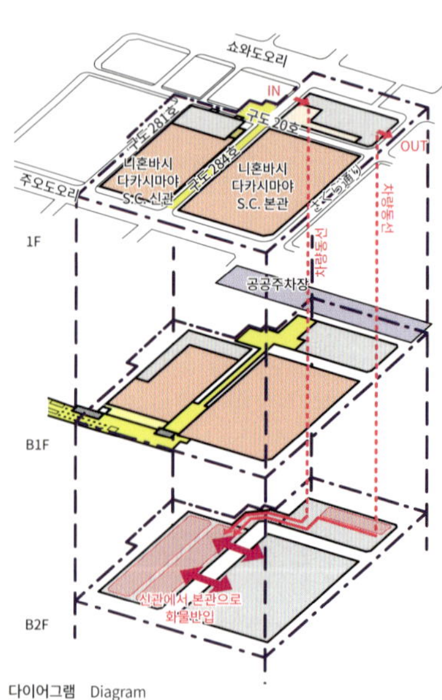

다이어그램 Diagram

여러 블록의 수화물 반출입 공간과 차량동선을 연계
Loading space and vehicle circulations coordinated across multiple city clocks.

니혼바시 갤러리아. 지역의 활기 계승을 위해 계획된 도로공간.
Nihonbashi Galleria. A street space aimed to inherit the liveliness of this area.
민간부지
Private Property
민간부지
Private Property
중앙구도(보행자전용도)
Chuo City Road(Pedestrian Walkway)

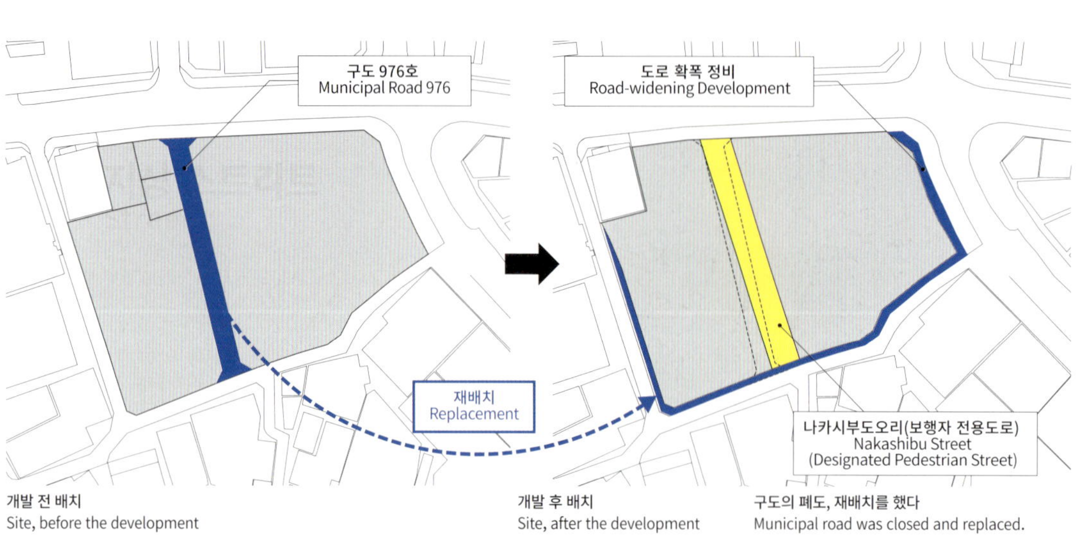

남쪽에서 바라봄. 왼쪽은 상점가로 활력 넘치는 시부야 PARCO 부지 내의 나카시부 도오리 (보행자 전용 도로)
View from the south. On the left is a through passage on the site of Shibuya PARCO full of liveliness (Designated Pedestrian Street).

SEAMLESS LINKAGE

개발 전 배치
Site, before the development

개발 후 배치
Site, after the development

구도의 폐도, 재배치를 했다
Municipal road was closed and replaced.

Continuing the town's liveliness by replacing a closed road

— **Shibuya PARCO·HULIC building   Nakashibu Street**

폐도/재연결로 활력을 연결하다

시부야 파르코 휴릭빌딩, 나카시부 도오리(2019)

시부야 PARCO 재건축과 함께 탄생한 퍼블릭 스페이스. 종래의 구도(区道)를 폐도하고 부지 주변을 재정비해 상부에 상업시설을 만들면서도 저층부는 보행자 전용도로인 나카시부 도오리를 정비하는 것으로 종래의 보행자 동선을 확보하고 있다. 부지 남쪽의 스페인쟈카와의 연계성을 고려해 동선의 위치가 수정되어 보다 연속적인 보행자 네트워크를 실현했다. 또한 기능면에서도 상업의 활기를 연속시키기 위해 나카시부 도오리쪽에 연도상업을 배치했다.

남쪽에서 종전의 구도 976호를 바라봄
View of the former Municipal Road 976 from the south.

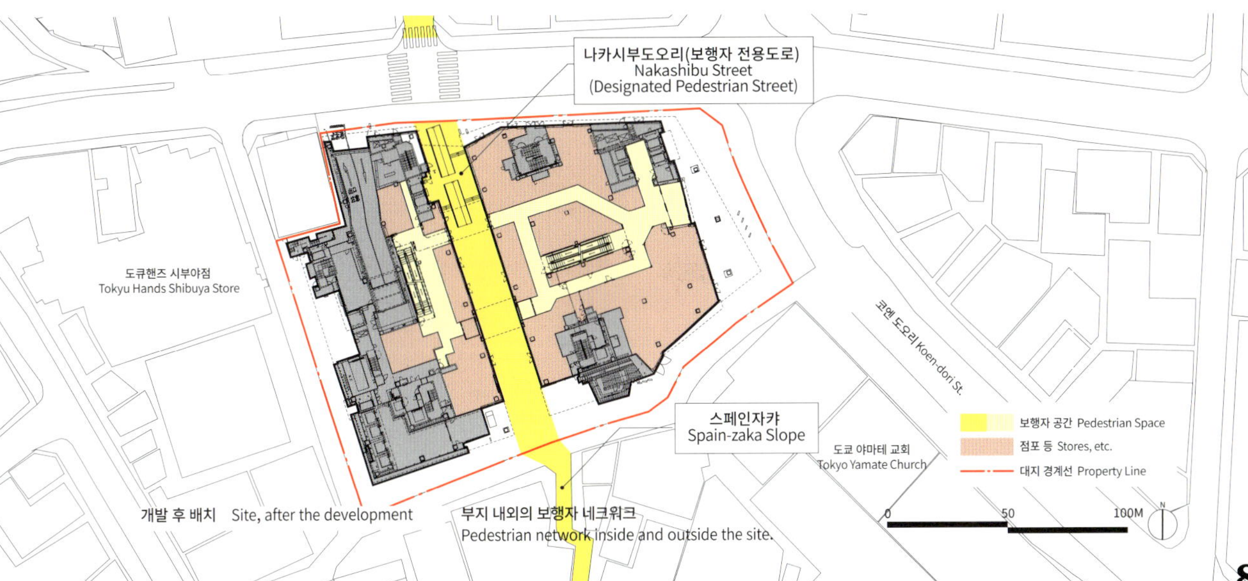

개발 후 배치   Site, after the development

부지 내외의 보행자 네크워크
Pedestrian network inside and outside the site.

Succeeding the town layout from Edo

TOKYO SQUARE GARDEN　Through Passage
KYOBASHI EDOGRAND　Through Passage

에도시대부터 지속된
구획나누기의 계승

도쿄 스퀘어가든 부지 내 공공통로(2013)
교바시 에도그랑 부지 내 공공통로(2016)

도쿄 스퀘어가든과 교바시 에도그랑에서는 개발 당시 기존의 구도를 폐지한 후 재연결 혹은 내부 도로를 만들어 부지 내에 주변 도로와 연결되는 공공통로를 정비했다. 통로와 접한 부분에는 점포와 오피스 로비 등을 배치하여 에도시대부터 지속된 구획을 계승함과 동시에 현대적인 퍼블릭 스페이스로 새롭게 태어났다. 도쿄 스퀘어 가든은 건물 저층부의 풍부한 녹지가 보행자를 맞이하는 쾌적한 보행통로를 만들어냈다. 교바시 에도그랑에는 점포뿐만 아니라 벤치 등을 배치해 쉼터와 같은 역할을 부여하여 사람들이 자유로이 이용할 수 있는 공간을 만들었다. 두 프로젝트는 이처럼 비슷한 정비 수법을 사용하면서도 서로 다른 퍼블릭 스페이스를 저층부에 전개하고 있다.

도쿄 스퀘어가든 부지 내 공공통로
Through passage of TOKYO SQUARE GARDEN.

교바시 에도그랑 부지 내 공공통로
Through passage of KYOBASHI EDOGRAND.

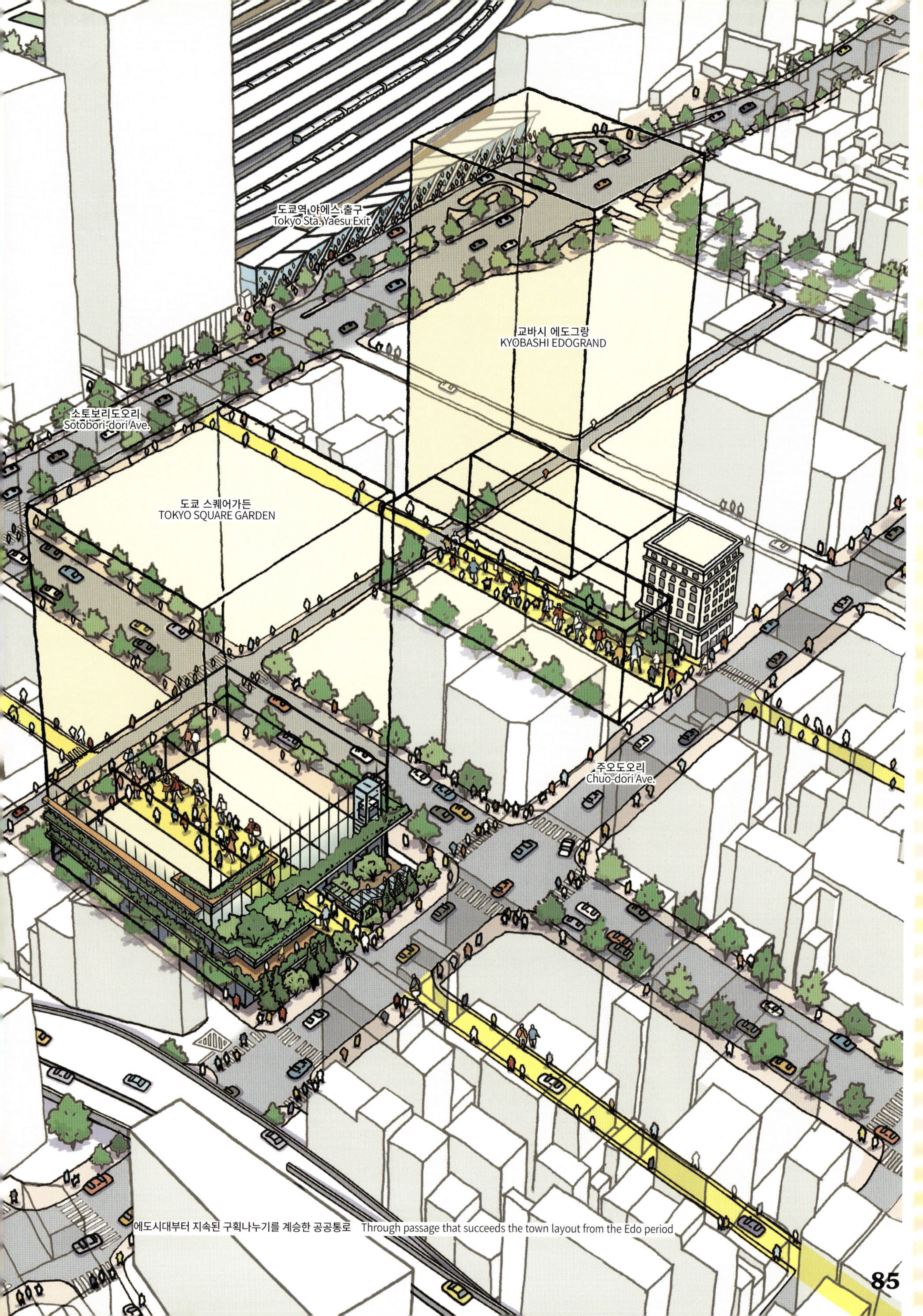

에도시대부터 지속된 구획나누기를 계승한 공공통로　Through passage that succeeds the town layout from the Edo period

*Pedestrian network formation process
guided by high-level planning*

도시계획이 유도하는 보행자 네트워크 형성의 프로세스

이 장에서 다루고 있는 것처럼 광범위한 보행자 네트워크를 형성하기 위해서는 다수의 개발자와 도로관리자, 행정기관, 또는 역이 인접한 경우에는 철도사업자를 포함한 다양한 Stake holder가 공공의 비전을 가지고 도시개발을 추진해 각각의 개발사업 중에 보행자 공간을 정비해 갈 필요가 있다.

이를 위해서 중요한 것은 마스터 플랜과 가이드 라인이다. 즉 행정에서 주도하여 개발의 내용을 계획하고 민관일체로 만드는 보행자 네트워크 방침을 결정한다. 개발사업자와 도로관리자, 철도사업자는 이를 근거로 보행자 네트워크의 상세계획을 검토한다.

각 개발의 계획검토가 심도 있게 진행됨에 따라 상위계획의 표현도 구체화된다. 도시계획 결정을 동반하는 개발에서는 보행자 공간의 위치와 크기를 도시계획상에 자리매김시키고 (예를 들면 지구계획상의 지구시설 등), 민간개발 내의 공공공간 정비를 담보한다.

일례로 시부야역 중심지구에 있어서 보행자 네트워크 형성의 프로세스를 소개한다. 여러 블록의 개발이 얽혀 10년이 넘는 장기간 동안 정비를 위해 개발의 진행에 맞춰 다양한 상위계획이 만들어지고 있다.

COLUMN

2010

시부야 중심지구 도시만들기 방침
여러 블록의 개발 구상을 근거로 보행자 네트워크의 큰 방향성을 결정한다.

■역 중심 지구 보행자 네트워크 도면 (데크) : 골짜기 형태의 역 중심 지구를 수평으로 연결

① 도켄자카 위에서 미야마스자카 위를 데크 레벨로 연결한다
② 국도 246호 쪽에 JR 노선 위를 연결하는 역 동쪽 출구, 서쪽 출구를 연결
③ 4층 데크 공간 정비 및 JR 상공을 남북으로 연결
④ 역 서쪽 출구를 남북으로 연결
⑤ 역 동쪽 출구를 남북으로 연결
⑥ 국도 246호 횡단 데크를 갱신
⑦ 역 서쪽 출구를 동서로 연결
⑧ 역 동쪽 출구를 동서로 연결
⑨ JR 신 남쪽 출구까지 남북으로 연결
⑩ JR 노선을 동서로 연결
⑪ 국도 246호에서 다이칸야마 방면으로 연결
⑫ 국도 256호에서 에비스 방면으로 연결
⑬ JR노선에 신 남쪽 출구와 사쿠라가오카 방면을 연결

2012

시부야역 중심지구 기본 정비 방침
도시만들기 방침을 근거로 각 지구의 보행자
네트워크 계획 검토를 심도화

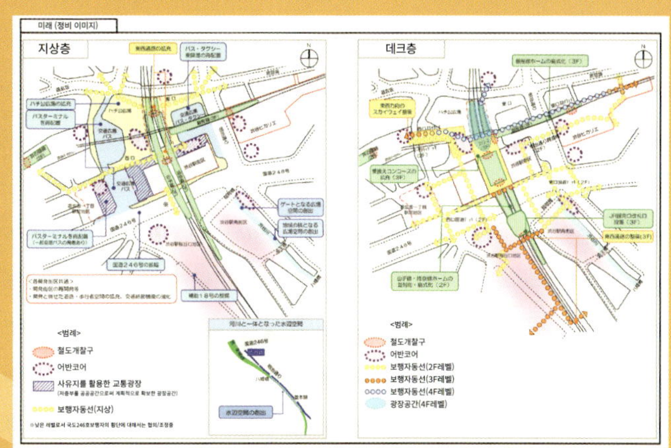

2013

시부야 3초메지구 지구계획
계획이 더욱이 심도화되어 보행자 공간의 구
체적이 위치와 크기를 개별지구의 도시계획에
서 결정. 여기에서는 예로써 시부야 스트림에
관련한 지구계획만을 개재하고 있다.

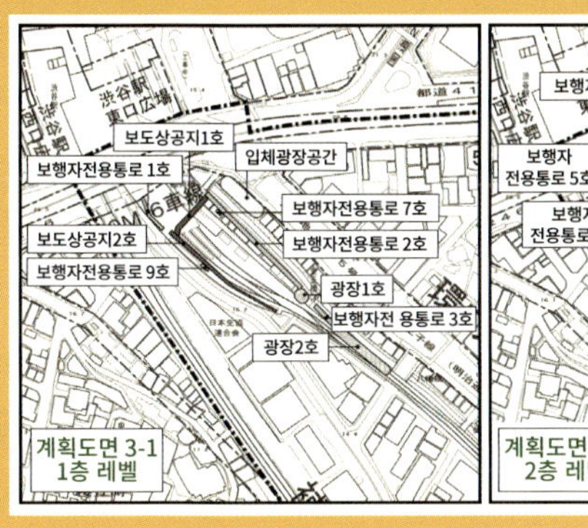

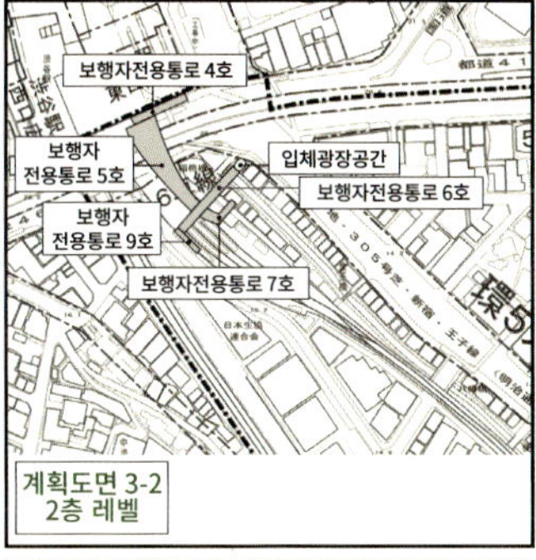

2020

2020년 8월 시점의 보행자 네트워크
도시계획 결정을 통해 각 지구에서 상세설계
및 건설 진행. 보행자 공간이 순차적으로 실현
되고 있다.

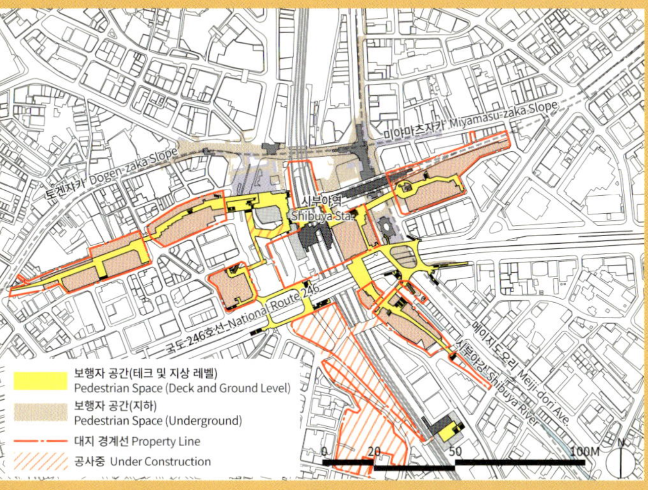

Underground corridors that support the heart of Tokyo

Otemachi-Marunouchi-Yurakucho District　Underground Pedestrian Network

도쿄 중심부의 회유성을 만들어내는 지하 회랑

다이마루유 지구 보행자 네트워크

오테마치, 마루노우치, 유락초 지구 도시만들기 가이드라인을 근거로 정비된 지하 보행자 네트워크. 골격이 되는 다이묘 코지를 중심으로 각 지구에서 민간개발에 의한 보완적인 지하 보행자 네트워크가 정비되는 것으로, 지하철 5개 노선을 중심으로 한 지하 회랑이 구성되었다.

도영 미타선 오테마치역
Toei Mita Line Otemachi Sta.
치요다선 오테마치역
Chiyoda Line Otemachi Sta.
고쿄 도오리
Gyoko-dori Ave.
한조몬선 오테마치역
Hanzomon Line Otemachi Sta.
마루노우치 빌딩
Marunouchi Bldg.
신 마루노우치 빌딩
Shin-Marunouchi Bldg.
오테마치 타워
The Otemachi Tower
마루노우치선 도쿄역
Marunouchi Line Tokyo Sta.
토자이선 오테마치역
Tozai Line Otemachi Sta.
마루노우치선 오테마치역
Marunouchi Line Otemachi Sta.
정비 예정
Development Schedule
오테마치 플레이스
Otemachi PLACE
JR 도쿄역
JR Tokyo Sta.
한조몬선 미츠코시마에역
Hanzomon Line Mitsukoshimae Sta.
지하 보행자 공간
Underground Pedestrian Space
블록 라인 City Block Line
토자이선 니혼바시역
Tozai Line Nihombashi Sta.
0   40   100   200M
배치 Site

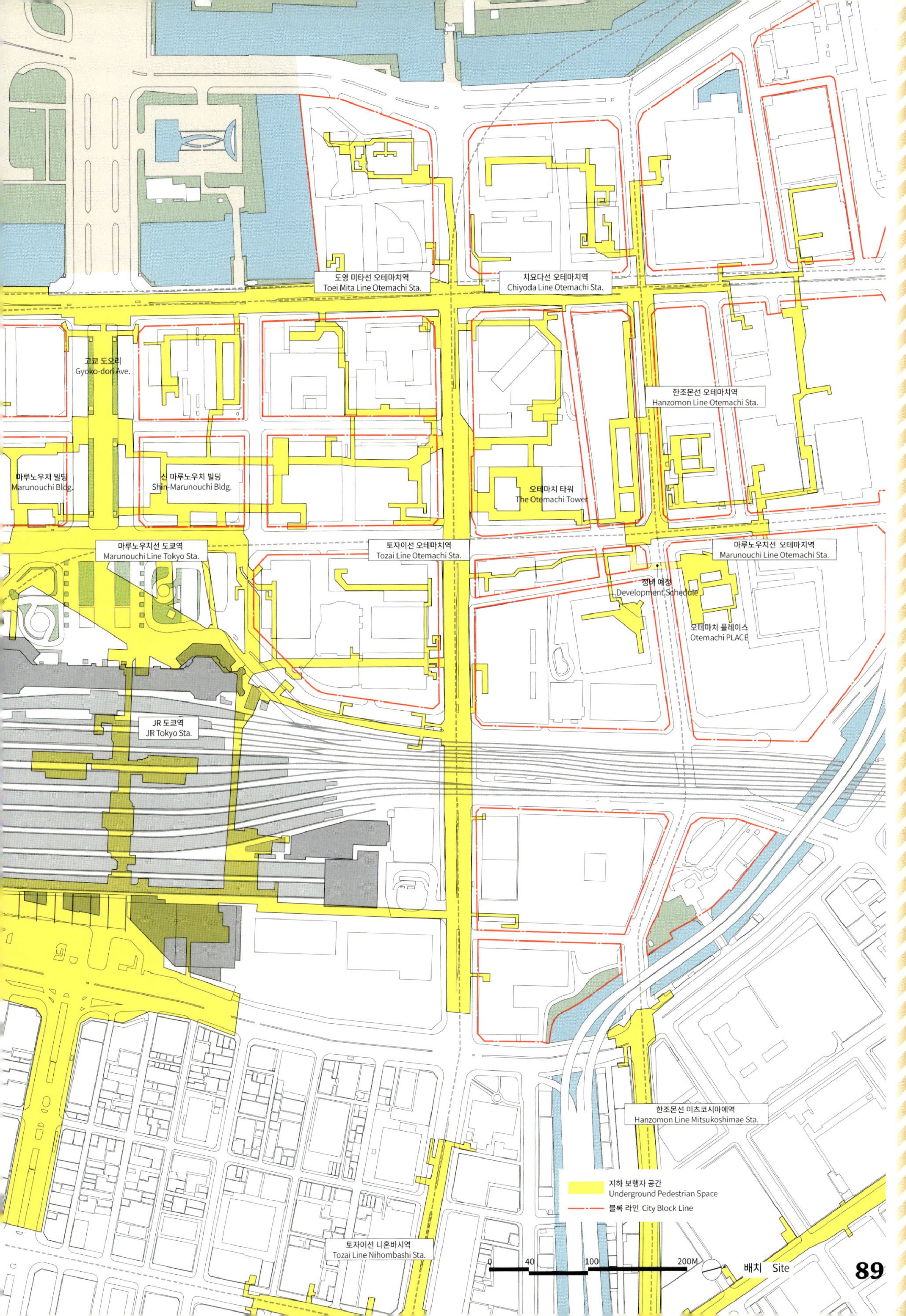

89

MAP p.189

21

SEAMLESS LINKAGE

원활한 네트워크를 형성하는 퍼블릭 스페이스

지하철 환승 거점이 되는 오테마치 타워 지하 2층 플라자
Plaza on the 2nd basement floor of The Otemachi Tower
functions as subway transfer hub.

민간부지 내에 만들어진 지하철 환승 거점

오테마치 타워 프라자(2014)

오테마치 타워에서는 도시재생특별지구를 활용하여 다이마루유 지역의 거대한 지하 보행자 네트워크를 구성하는 것의 일환으로써 지하철 5개 노선의 환승 거점이 만들어졌다. 특히 도쿄메트로의 정식적인 환승 동선인 부지 동쪽의 토자이선/마루노우치선의 지하연락통로의 폭을 4m에서 8m로 확대하여 「프라자」와 일체로 정비함으로써 자연광이 충만한 알기 쉬운 지하 환승 동선을 실현했다. 더욱이 「프라자」에서 지상의 「오테마치의 숲」으로 연결되는 4개층분의 오픈공간은 지하에서 지상으로 동선적, 시각적으로 연속된 중간영역을 만들고 있다.

Connecting the Otemachi-Marunouchi-Yurakucho District and surrounding areas

**Otemachi Place
Sunken Garden, Central Promenade, Ryukan-sakura Bridge**

다이마루유와 주변지역을 연결한다

오테마치 플레이스
선큰가든 센트럴 프롬나드 류제키사쿠라교(2018)

오테마치 플레이스는 비지니스 중심지에 있으면서도 저층부를 도시에 개방한 공공공간이 되어 주변지역으로 사람들의 흐름이 새로 생겨났다. 지하철 오테마치역에 접속 예정인 선큰가든과 건물 안을 복층에 걸쳐 관통하는 「센트럴 프롬나드」를 설치해 지하, 지상, 2층 레벨의 입체적인 보행자 네트워크를 형성하고 있다. 센트럴 프롬나드에는 누구든지 이용할 수 있는 쾌적한 쉼터가 곳곳에 설치되어 비즈니스 워커들이 제3의 업무 공간으로 이용하고 있다.
2층 레벨은 부지를 넘어 니혼바시강을 건너는 류제키사쿠라교에 접속해 칸다, 니혼바시 방면으로 사람들의 왕래를 만들고 있다. 또한 교각의 종착점은 오테마치강의 녹도(綠道)와 연결되어 강을 따라 도시의 회유성을 높이고 있다.

사람들의 쉼터가 되는 선큰가든
Sunken Garden as a place for recreation.

부지 내를 관통하는 센트럴 프롬나드를 내려다본 관경
Looking down at Central Promenade cutting through the site.

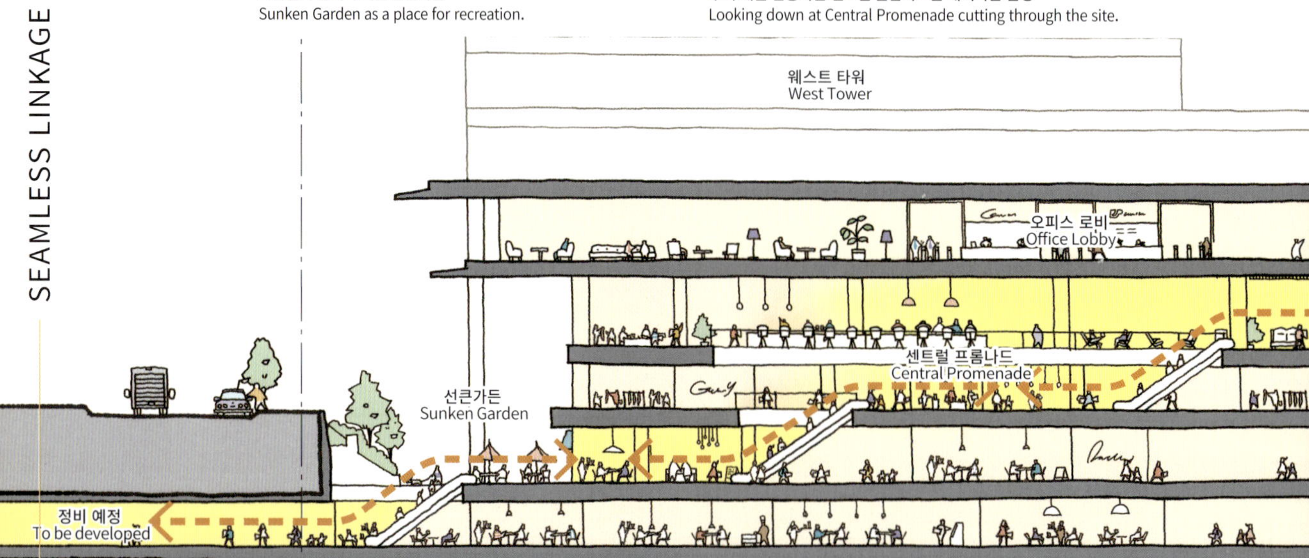

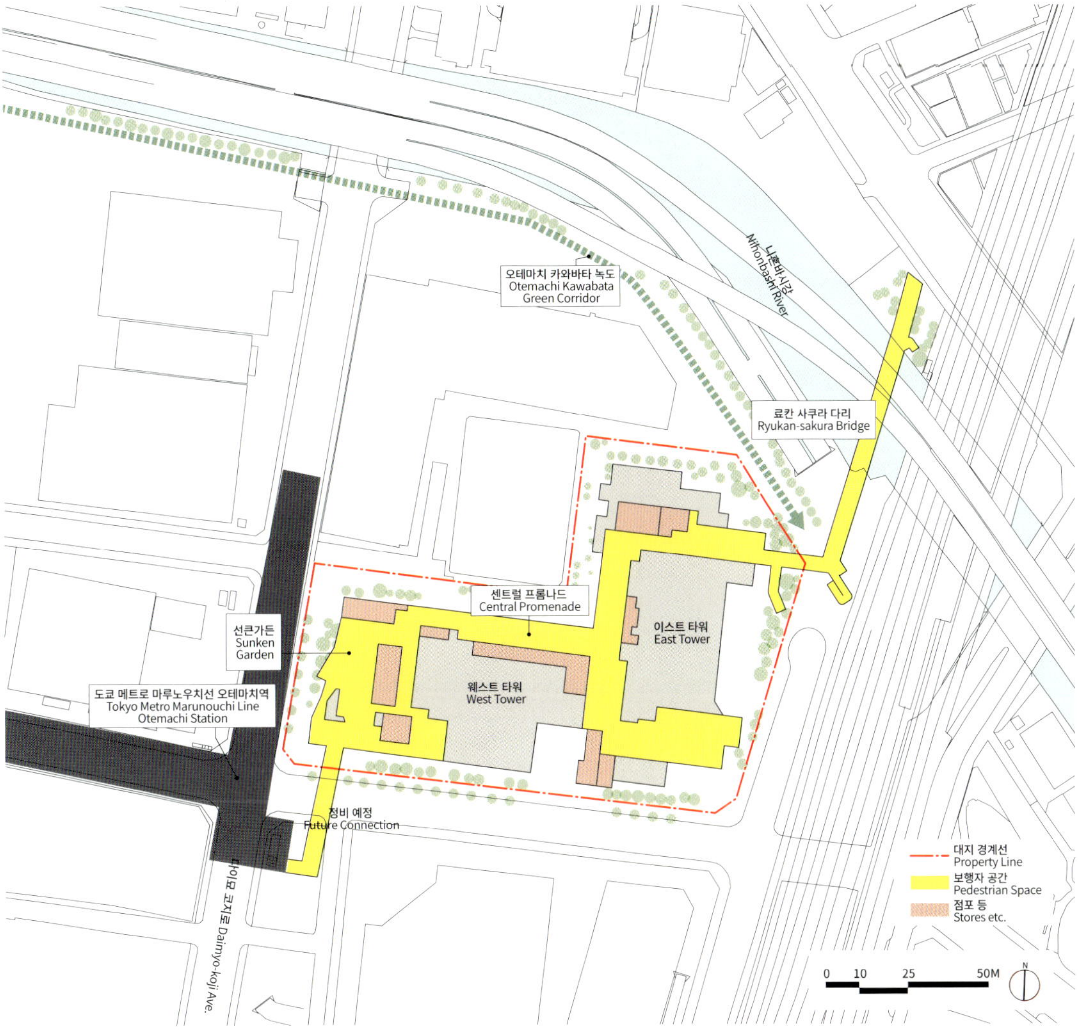

배치   Site

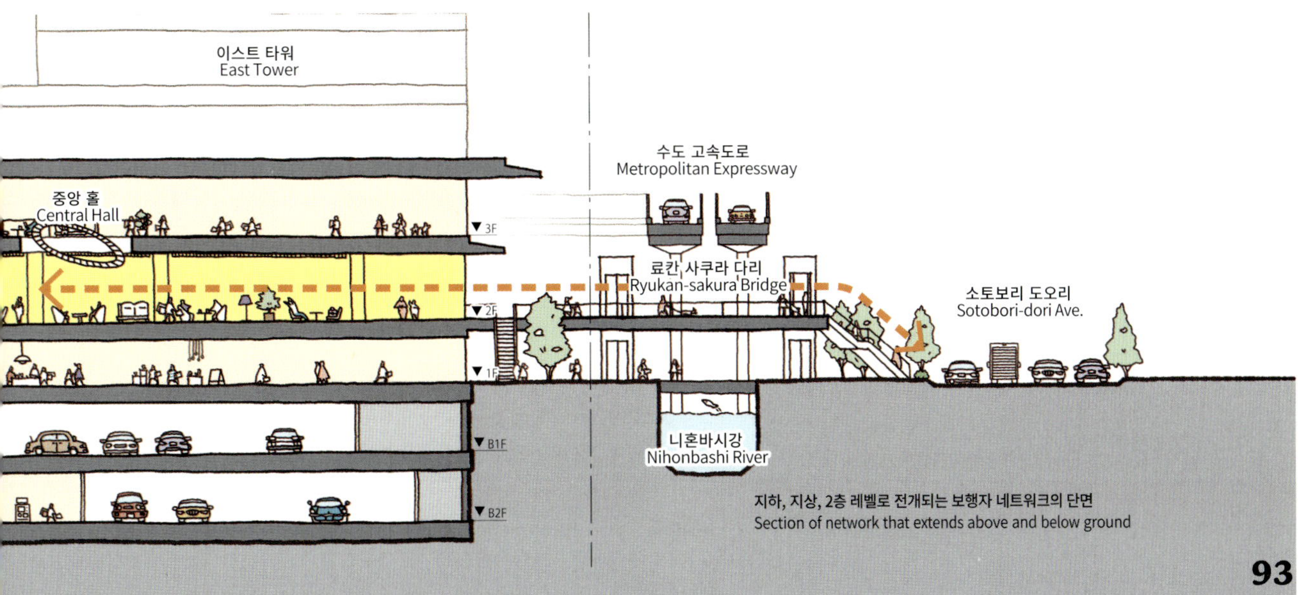

지하, 지상, 2층 레벨로 전개되는 보행자 네트워크의 단면
Section of network that extends above and below ground

Underground plaza that combines the underground walkway and sunken plaza

— Shiodome Sio-site   Underground Walkway and Sunken Plaza

지하보도와 선큰광장이 일체화된 지하광장

시오도메 사이트 지하보도와 선큰광장(1994 ～ 2011)

신바시역에서 시오도메역을 연결하는 공공지하보도에 접한 각 민간개발 부지 내에는 세 곳의 선큰광장이 정비되어 있다. 지하보도와 선큰광장이 일체로 연결됨으로써 자연광이 스며드는 지하광장이 형성되었다. 또한 도로 구역으로 되어 있는 지하보도는 도로점용 허가를 받은 편의시설이 설치되어 지하 보행공간임에도 불구하고 상업의 활기가 더해진 쾌적한 공공공간이 되었다.

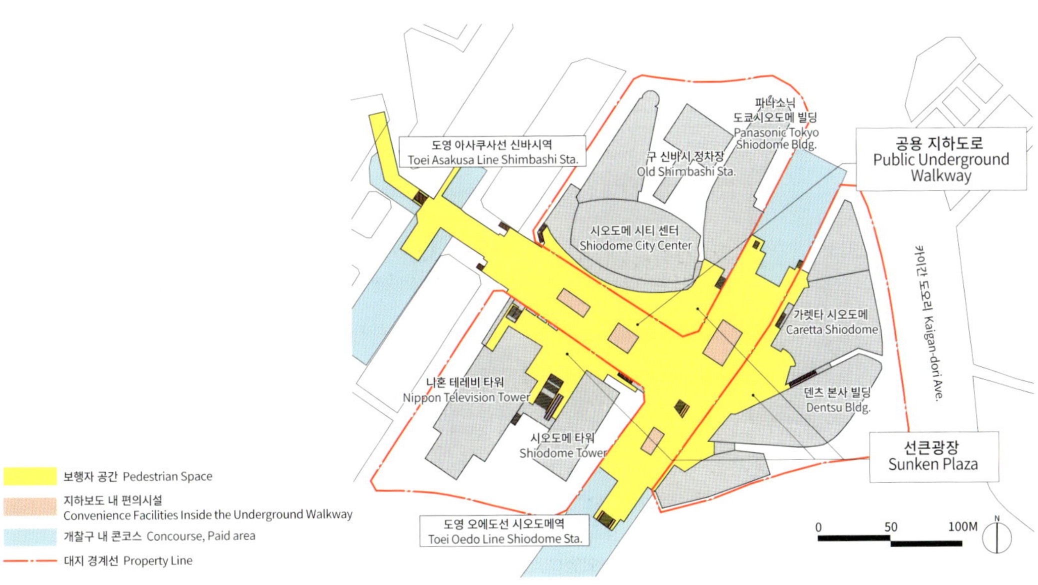

시오도메 사이트 공공지하보도. 안쪽으로 선큰 광장이 보인다.
Public Underground Walkway at Shiodome Sio-site. Sunken plaza can be seen at the back.

지하평면 Basement floor plan

Three-dimensional green corridor connected through time

Roppongi 1-chome Station Network

시간의 경과와 함께 연결되는 녹지의 입체 회랑

롯본기1초메역 주변 네트워크

아크힐즈(1986), 시로야마 가든/카미야초(1991), 롯본기1초메역(2000),
이즈미가든 타워(2002), 아크힐즈 사우스타워(2013)

지하철 신역(롯본기1초메역)과 일체적으로 계획된 역 동쪽 출구의 이즈미가든 재개발은 콘코스와 역 앞 광장 등 역 기능의 일부를 재개발 범위에 포함시켜 정비해 지하와 지상의 약 20m의 고저차를 현대의 새로운 언덕길 「Urban Corridor」와 연결시켰다. 그후 역 서쪽의 롯본기 그랑타워의 재개발로 시가지의 변신에 맞춰 역기능이 강화되고 확충됨으로써 도시의 골격이 되는 동서의 보행자 네트워크가 형성되었다. 또 이들의 보행자 공간은 아크힐즈의 카라얀광장과 이즈미도오리, 카미야초 녹도 등 주변의 퍼블릭 스페이스에도 연결되고 있다. 풍부한 녹지와 함께 입체적인 보행자 회랑은 여러 개발의 단계적인 정비로 장기간에 걸쳐 도시를 연결하고 있다.

롯본기 그랑타워 내의 지하철 역 앞 광장에서 액세스 몰을 바라본 관경
View of Access Mall from the subway station plaza at Roppongi Grand Tower.

롯본기1초메역 동쪽 출구 개찰구에서 이즈미가든과 Urban Corridor를 바라본 관경
View toward Izumi Garden Tower's Urban Corridor from the East Gate of Roppongi 1-chome Station.

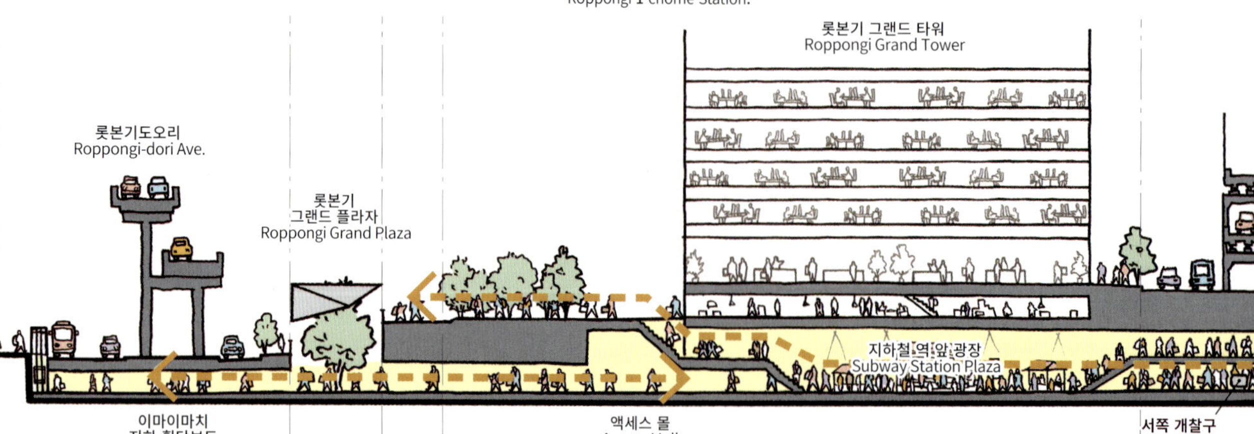

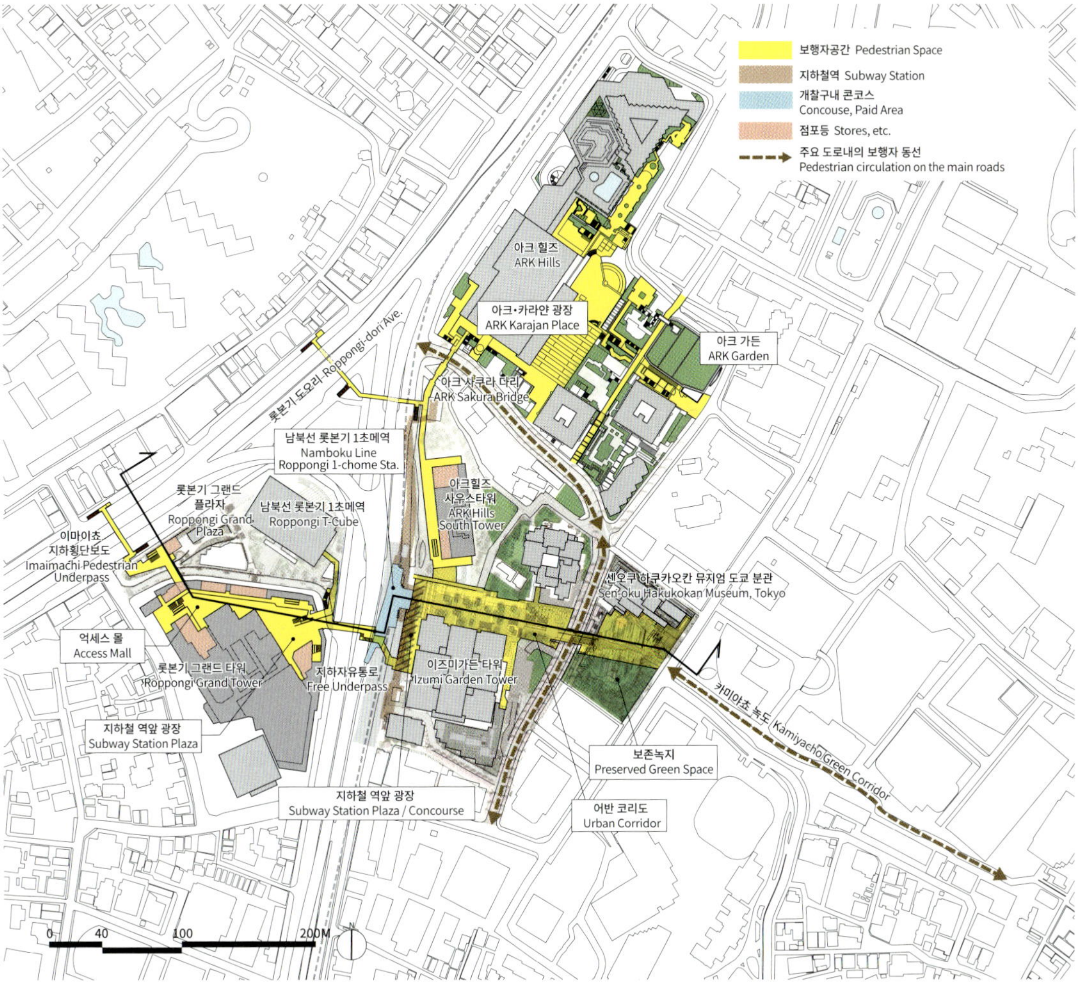

배치　Site　롯본기 1초메역 주변에는 단계적으로 개발된 복수의 부지가 보행자 네트워크로 연결된다.
In the area around Roppongi 1-chome Station, multiple sites developed in stages are connected by a pedestrian network.

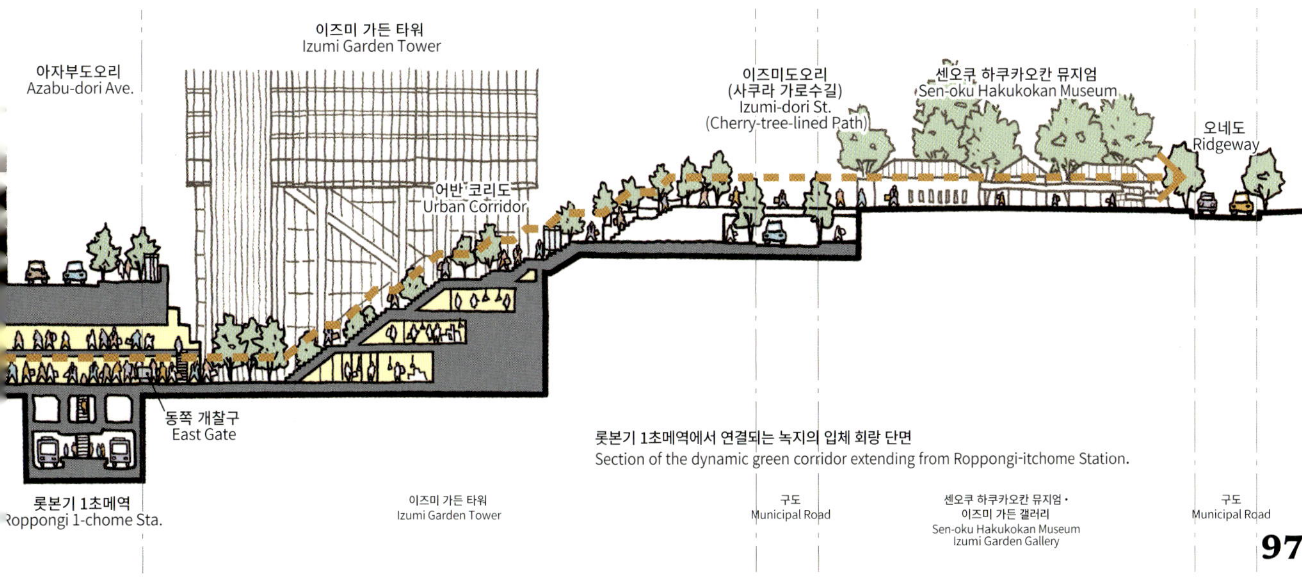

롯본기 1초메역에서 연결되는 녹지의 입체 회랑 단면
Section of the dynamic green corridor extending from Roppongi-itchome Station.

Subway becomes a starting hub of the local network through a Station Plaza on private property
— Network around Toranomon Hills Station

▶ ANTICIPATED PROJECT

민간용지 내의 역광장을 기점으로 한 네트워크로 지하철 결절점을 만들다

토라노몬힐즈역 주변의 네트워크
토라노몬힐즈역(2020), (가칭)토라노몬힐즈 스테이션타워(2023)

도쿄메트로 히비야선 토라노몬힐즈역은 인접한 토라노몬힐즈 스테이션타워의 개발지와 함께 계획되어 민간용지 내에 역과 도시를 연결하는 입체적인 역광장이 정비된다. 종래의 도로하부라고 하는 공간적 제약이 큰 지하철역이지만 지금까지 전례가 없었던 밝고 넓은 지하 퍼블릭 스페이스가 만들어질 것이다.

아카사카/토라노몬 에리어는 여러 민간개발과 연계하여 지하, 지상, 데크레벨의 보행자 네트워크가 정비되고, 토라노몬역, 토라노몬힐즈역, 다메이케산노역의 세 역이 연결될 예정이다. 복합적인 기능집적이 진행되는 토라노몬 지구 전체의 회유성을 높이는 것과 동시에 지하철 노선 간의 환승이 가능한 지하철 결절점의 창출도 기대되고 있다.

실내 보행자 통로의 완성 이미지
Completion image of internal pedestrian walkway.

민간용지 내의 역광장 완성 이미지
Completion image of the station plaza on private property.

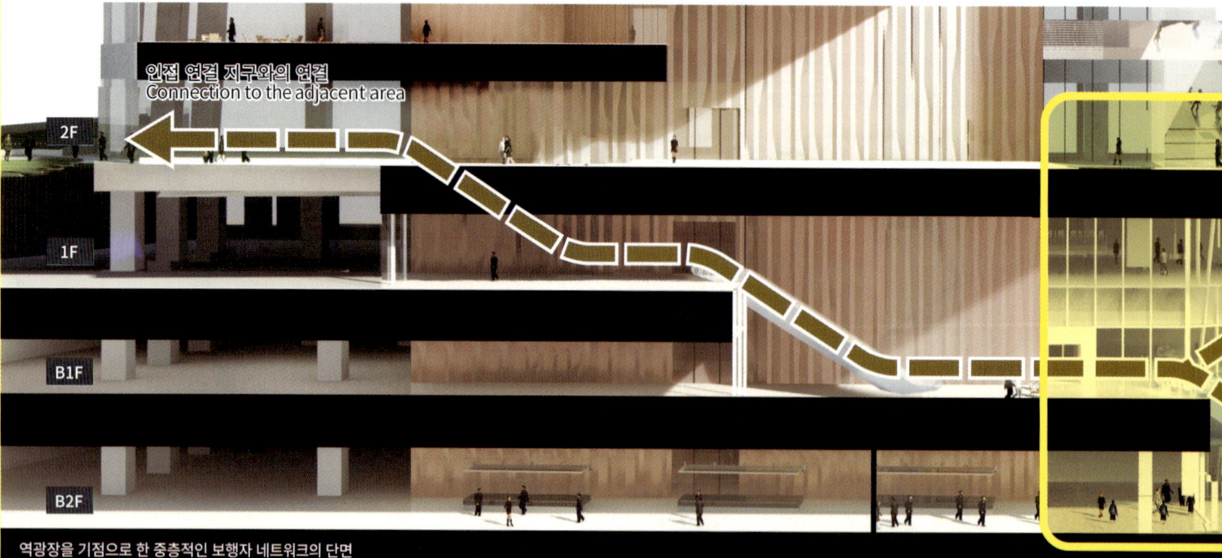

역광장을 기점으로 한 중층적인 보행자 네트워크의 단면
Section of multi-layered pedestrian network starting at the station plaza

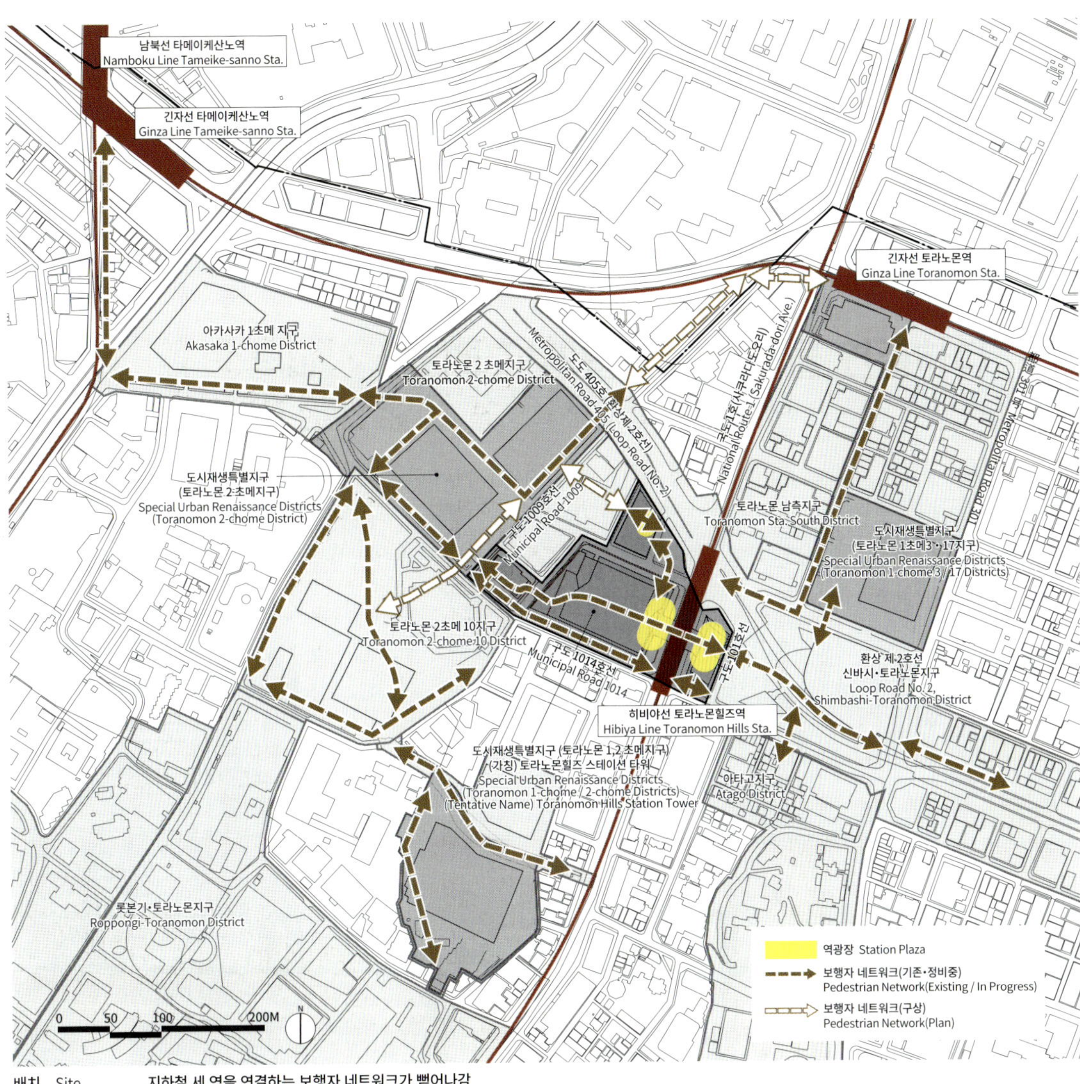

배치　Site　지하철 세 역을 연결하는 보행자 네트워크가 뻗어나감
Pedestrian network connecting three subway stations extends across the area.

URBAN GREEN CORE

Key public spaces integrated
into the surrounding city

주변의 지역과 연계된
퍼블릭 스페이스의 거점

미드타운·가든의 개방적인 잔디 광장
MIDTOWN GARDEN's open Grass Square.

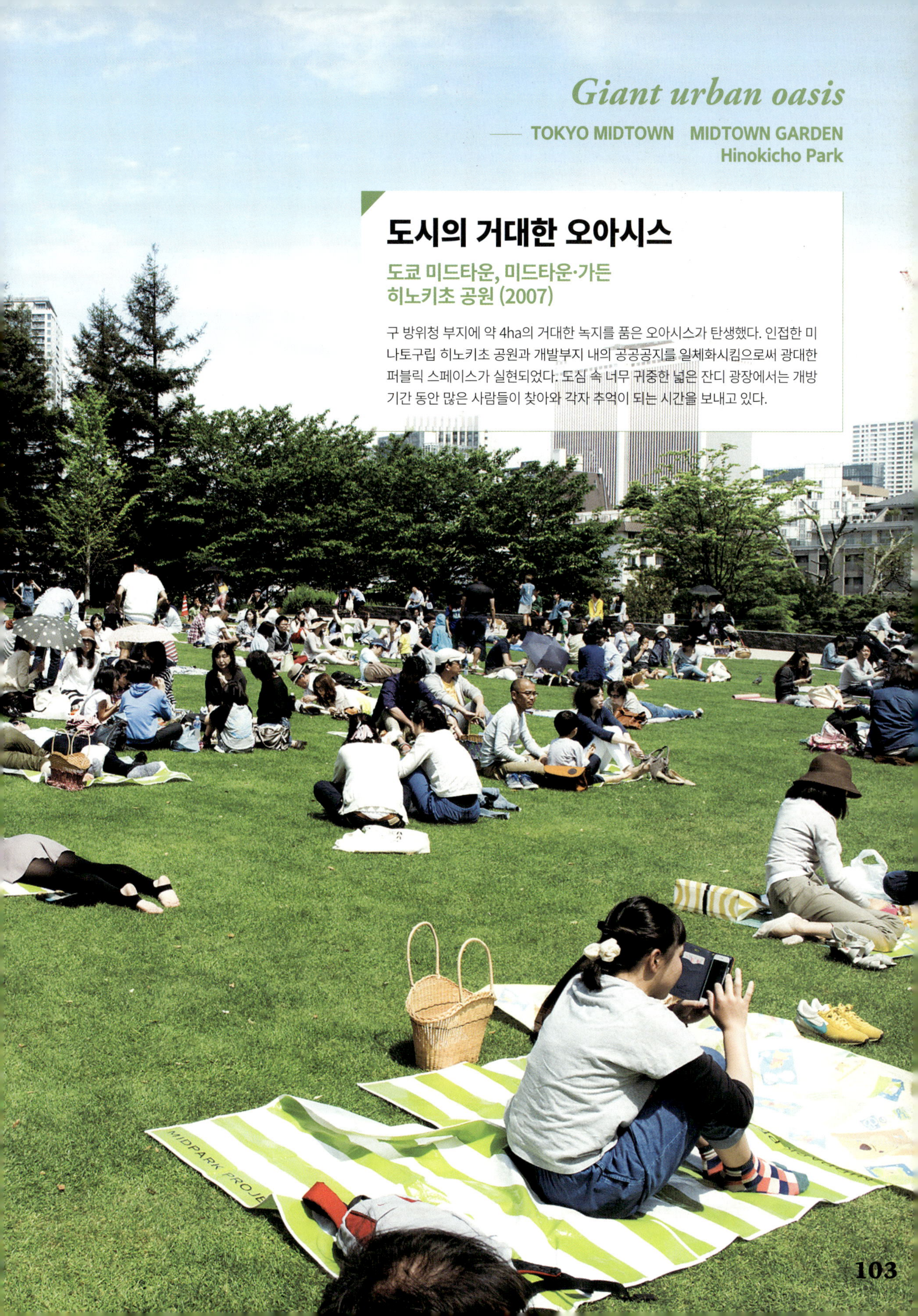

Giant urban oasis

— TOKYO MIDTOWN　MIDTOWN GARDEN
Hinokicho Park

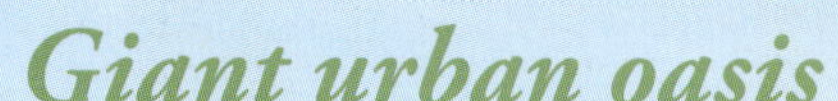

도시의 거대한 오아시스

도쿄 미드타운, 미드타운·가든
히노키초 공원 (2007)

구 방위청 부지에 약 4ha의 거대한 녹지를 품은 오아시스가 탄생했다. 인접한 미나토구립 히노키초 공원과 개발부지 내의 공공공지를 일체화시킴으로써 광대한 퍼블릭 스페이스가 실현되었다. 도심 속 너무 귀중한 넓은 잔디 광장에서는 개방 기간 동안 많은 사람들이 찾아와 각자 추억이 되는 시간을 보내고 있다.

미드타운·가든과 일체적으로 정비된 히노키초 공원
Hinokicho Park developed integrally with MIDTOWN GARDEN.

A function shared between a park and privately owned public space

공원과 공공 공지의 역할 분담

미드타운 가든은 지구 계획상 지구(地区) 시설로써 민간 개발 부지 내에 탄생되어, 미나토구립 히노키초 공원과 일체 정비가 이뤄졌다. 미드타운 가든은 잔디 광장이 특징인 반면, 히노키초 공원에는 놀이기구와 일본 정원이 설치되는 등, 성격이 다른 다양한 공간이 공존하여 다양하게 시간을 보낼 수 있는 퍼블릭 스페이스가 만들어졌다.

미드타운 가든과 히노키초 공원이 일체적으로 정비되었다.
MIDTOWN GARDEN and Hinokicho Park were developed integrally.

Seamless connection with the surrounding area

주변 지역과의 원활한 연결

개발 부지와 주변 지역 사이에는 레벨차가 있지만, 지구(地区) 시설에 정해진 외주 보행로와 인접한 대지 사이를 기존의 녹지 보존과 경관을 배려하는 완충지대로 만들어 주변과의 연결을 강화하고 있다. 또한 추후 개발된 파크 코트 아카사카 히노키초 더 타워에는 지구 계획의 변경으로 인해 외주 보행자 전용 도로에 접하는 형태로 새로운 보행자 동선이 정비되어, 아카사카 방면으로의 접근성도 좋아졌다.

북쪽에서 파크 코트 아카사카 히노키초 더 타워의 저층부 공간을 바라본 관경
View of the base area of Park Court Akasaka Hinokicho The Tower from the north.

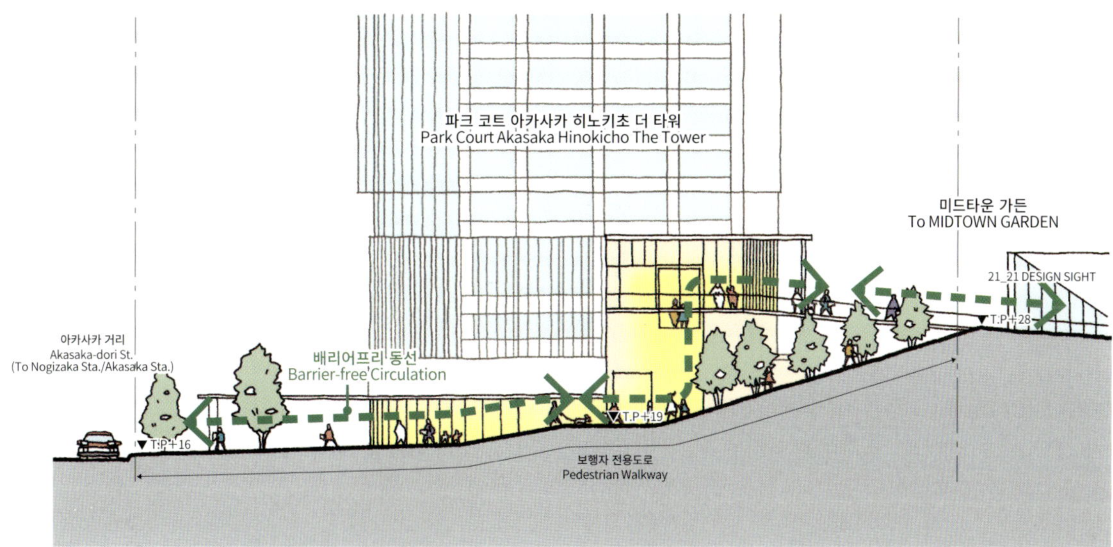

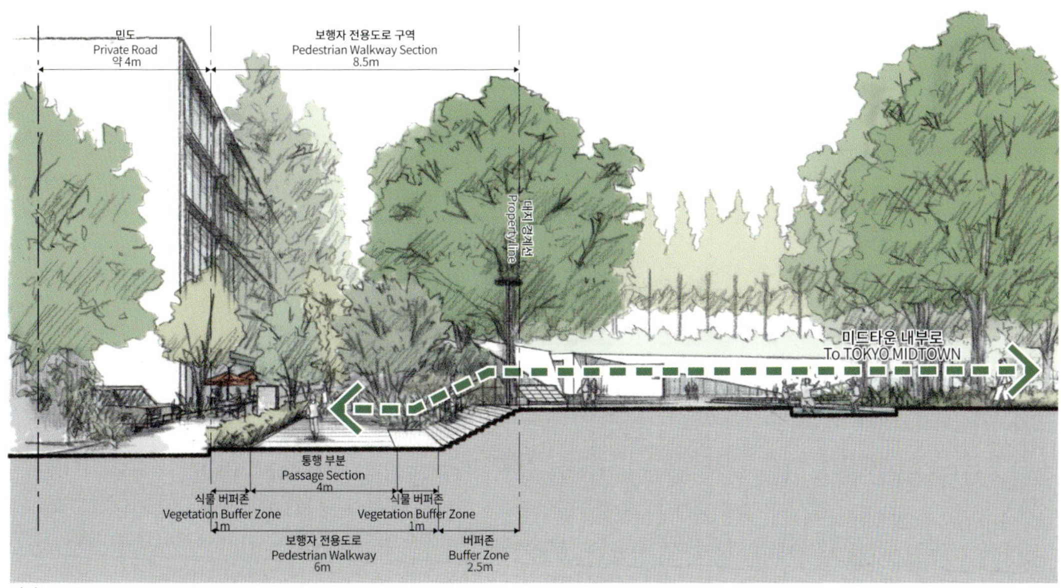

단면　Sectoin

공공 공지와 공원을 일체화한 녹지의 대공간

시나가와 센트럴가든(2003)

시나가와 인터시티(1998), 시나가와 그랜드 코몬(2003), 모리 공원 시오도메 공원

시나가와역 동쪽 출구에 정비된 「시나가와 센트럴가든」은 시나가와 인터시티와 시나가와 그랜드코몬 사이에 창출된 약 1.8ha의 거점적인 퍼블릭 스페이스다. 지역의 중앙을 관통하는 폭넓은 녹지의 대공간은 일체적인 공간으로 보이지만 사실은 여러 사업자의 공공 공지와 두 개의 공원으로 구성되어 있다. 이는 공공 시설과 민간부지의 경계가 없는 일체적인 공간으로 정비한 것으로, 녹지의 대공간은 시나가와역의 동쪽 출구를 상징하고 있다. 시나가와역과 연결되는 보행자 데크는 이 녹지를 따라가도록 배치되어 있으며, 각각에 있는 지상 데크 레벨은 사계절의 자연을 바라보면서 쾌적하게 이용할 수 있도록 만들어진 퍼블릭 스페이스다.

Grand green space integrating privately owned public space and parks

Shinagawa Central Garden

시오노 공원 (미나토구)
Shio Park (Minato City)

센트럴 가든 (공공 공지)
Central Garden (Privately Owned Public Space)

삼림 공원 (시나가와 구 · 미나토구)
Forest Park (Shinagawa City·Minato City)

남쪽에서 본 관경
View from the south

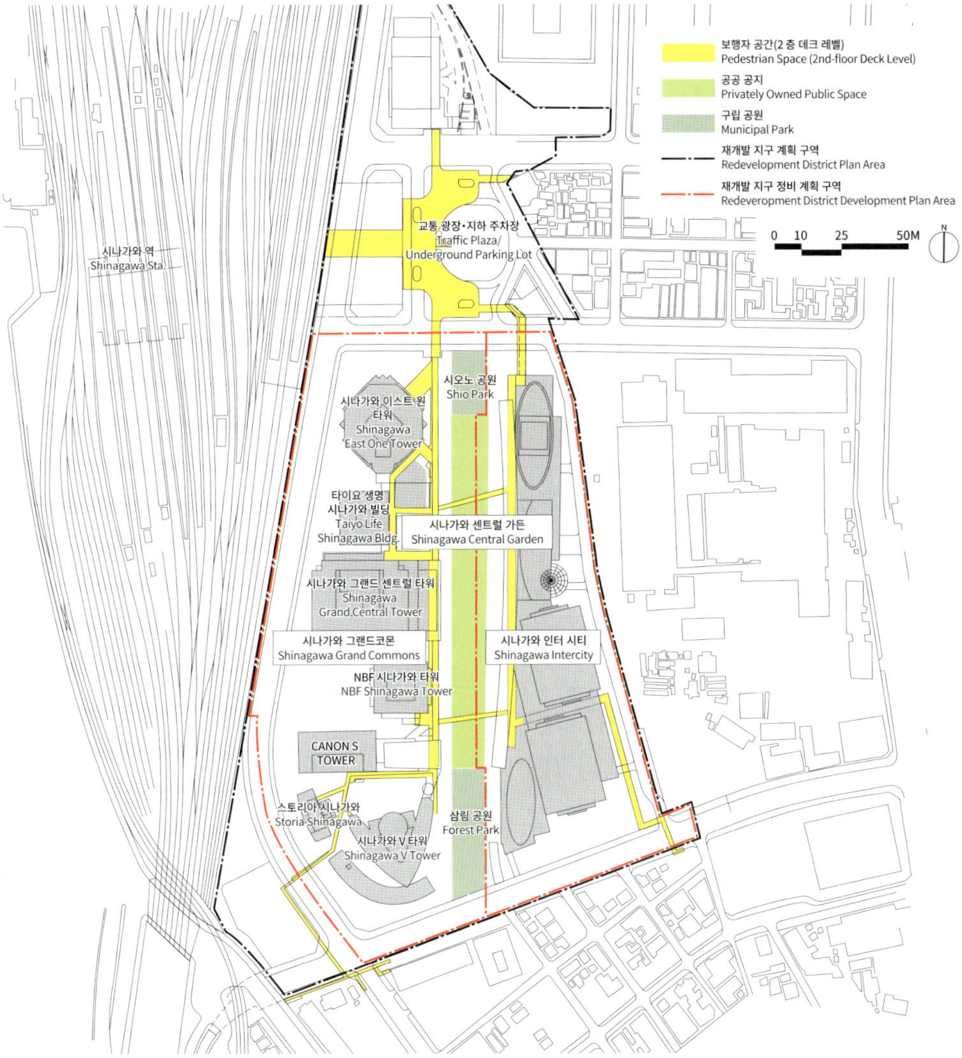

배치   Site

남쪽에서 본 관경. 사람들이 오고 가는 녹지의 대공간이 형성되어 있다.
View from the south. Grand green space occupied by people is created.

주변의 지역과 연계한
퍼블릭 스페이스의 거점
MAP 0.107
URBAN GREEN CORE
YELISU GARDEN 2019
108

다양한 이벤트가 개최되는 센터 광장 (사진은 이벤트 개최시)
Center Square where various events are held.

Surrounding the entire urban development with green to enrich pedestrian space

지역 전체를 녹음으로 둘러싸고, 풍부한 보행자 공간 확보

부지의 북쪽 폭 15m의 도시계획 도로 남쪽면에는 폭 5m의 보도상 공지로 인해 충분한 보행 공간을 확보했을 뿐만 아니라 큰 나무 2열 가로수길을 만들어 녹음이 풍부한 보행자 공간을 실현했다. 같은 가로수길로 보이지만, 한쪽은 도로의 가로수이고 다른 한쪽은 민간 부지 내의 가로수다. 거기에 더해 개발의 기부 공원으로서 도로와 일체적으로 아메리카 다리 공원이 정비되어, 개발과 주변 지역 사이의 녹음이 풍부한 보행자 공간을 형성하였으며, 이를 통해 부지 중앙의 센터 광장 등으로 자연스럽게 사람들을 맞이한다.

부지 북서쪽에 면한 T자 로를 바라본 관경. 공원, 도로, 민간 부지가 일체화된 녹음이 풍부한 보행자 공간
View toward T-junction facing the northwest side of the site. A park, roads, and private property are integrated to create pedestrian space full of greenery.

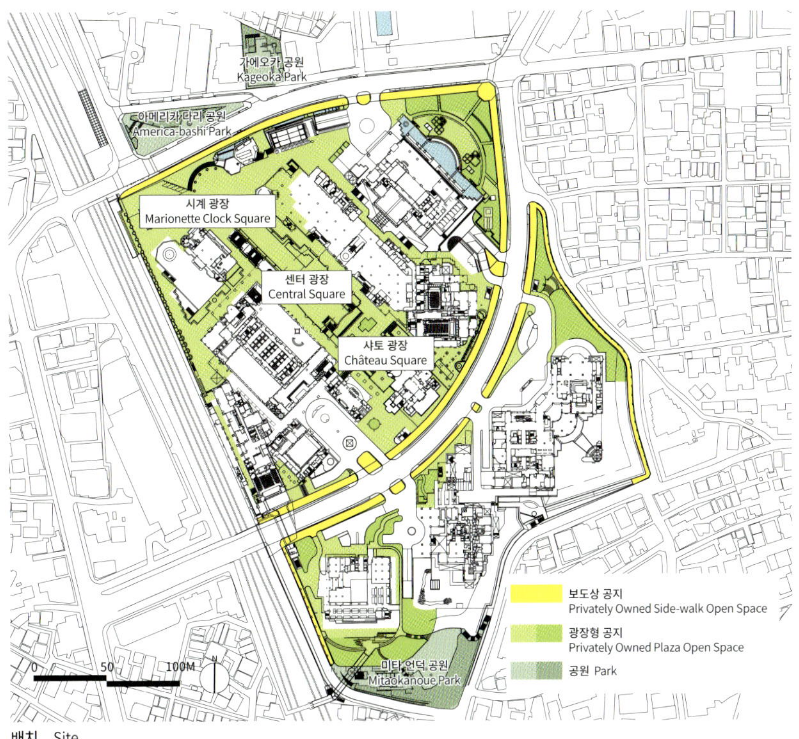

가에오카 공원
Kageoka Park
아메리카 다리 공원
America-bashi Park
시계 광장
Marionette Clock Square
센터 광장
Central Square
샤토 광장
Château Square
미타 언덕 공원
Mitaokanoue Park
0
50
100M
N
보도상 공지
Privately Owned Side-walk Open Space
광장형 공지
Privately Owned Plaza Open Space
공원 Park

배치 Site

민간 부지의 식재
Plantings in Private Property
가로수
Roadside tree
YEBISU GARDEN PLACE
5m 보도상 공지 (민간 부지)
5m Sidewalk Open Space (Private Property)
보도
Walkway
차로
Roadway
15m 도로 [보도 + 차도 + 보도] (시부야 구)
15m Road (Walkway + Roadway + Walkway) (Shibuya City)

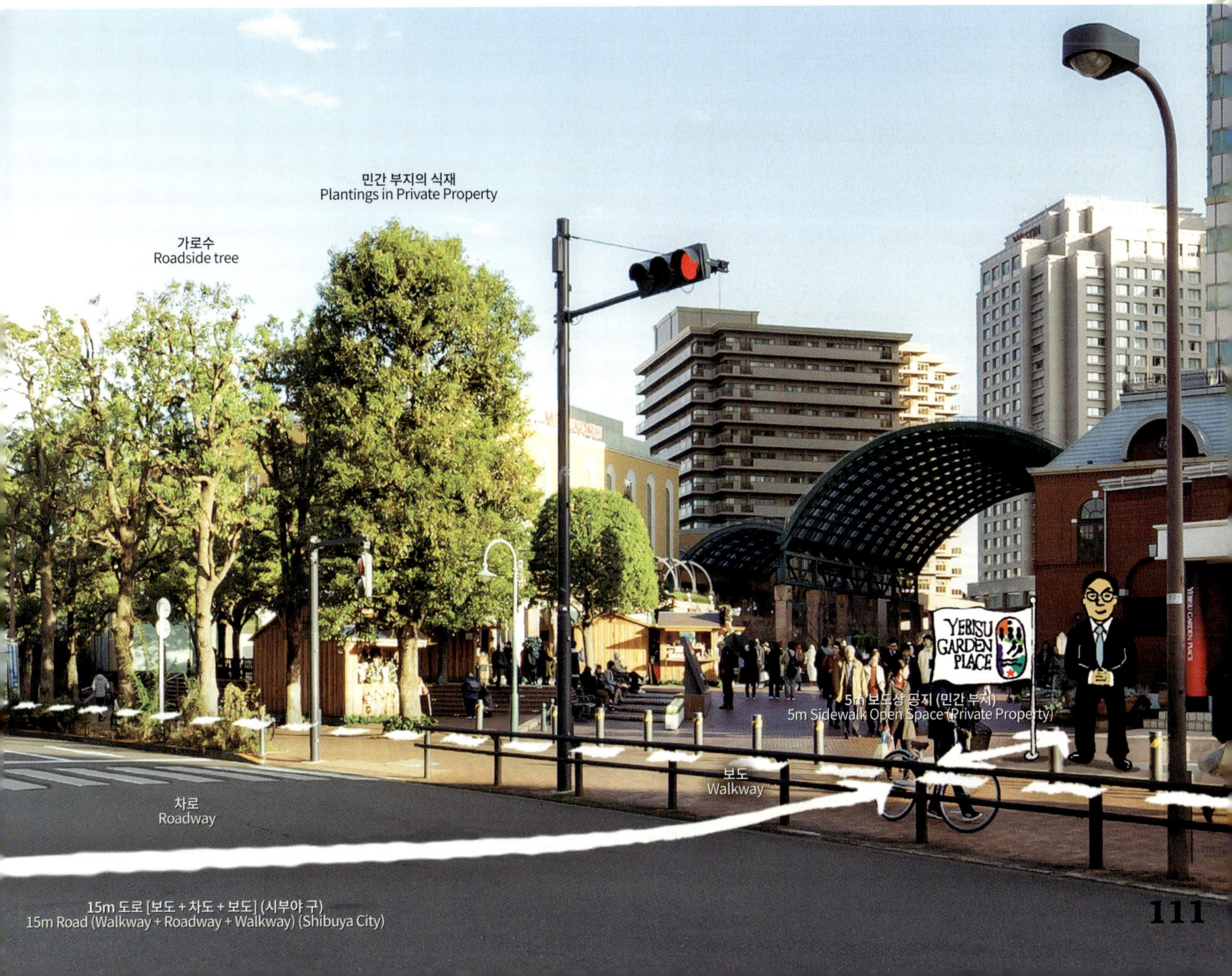

Two developments connected by a forest

ThinkPark   ThinkPark Forest
NBF Osaki Bldg.   Green Zone

두 개의 개발을 숲으로 연결

Think Park   Think Park Forest(2007)
NBF 오사키 빌딩 녹지(2011)

MAP p.194

서쪽 출구의 지구 개발을 유도하기 위해 책정된 오사키역 서쪽 출구 지구 디자인 가이드라인에서는 오피스 중심의 개발인 Think Park 및 NBF 오사키 빌딩 부지 서쪽에 직장인 및 주민의 휴식 환경의 축이 되는 「센터 그린」을 형성하기로 정하였다. 개발 시기도 개발 사업자도 다른 두 개의 개발이 가이드라인의 개념에 따라 지표면과 저층부에 건축과 일체된 풍부한 녹지를 조성함으로써 두 개발이 숲으로 연결되어 도쿄의 열섬현상 억제에 공헌하는 도시 스케일의 쿨 스폿을 실현하고 있다.

URBAN GREEN CORE

Think Park와 NBF 오사키 빌딩 부지 내에 형성된 센터 그린
Center Green created within the sites of ThinkPark and NBF Osaki Bldg.

북쪽에서 Think Park Forest를 본 관경
View of ThinkPark Forest from the north.

북쪽에서 NBF 오사키 빌딩 녹지를 본 관경
View of NBF Osaki Bldg. green zone from the north.

개발 부지 내에 창출된 히라카와의 풍경
Hirakawa no Michi created in the development site.

사람의 흐름을 받아들이는 나무그늘 공간

아이 가든 에어 히라카와의 풍경 외(2003)

아이 가든 에어는 이이다바시 화물역의 부지를 개발한 것이다. 화물역과 하천에 의해 분단되어 있던 동서 지역을 연결하기 위하여 개발 내의 기존 보행자 네트워크와 연계하는 보행자 동선을 치밀하게 정비하였다. 이 공간에는 녹색과 물이 풍부한 랜드 스케이프가 계획되어 있어, 주변을 걷는 사람들의 움직임을 부드럽게 받아들인다.

위: 프론트 포럼　아래: 가든 몰
Above: Front Forum. Below: Garden Mall.

The element that formerly divided the town now working as a connection

Plaza with continuous green penetrating through city blocks

한때의 분단 요소가 지역을 연결

지금까지 갈 수 없었던 니혼바시 강변에 벚꽃길 도로를 정비하면서 기존에 설치되어 있던 니혼바시 다리를 재건하였다. 도로변에는 보도형 공지·광장형 공지, 건물 사이에는 공공통로를 배치하여 주변의 보행자 네트워크와 연결시키고 있다.

이러한 계획으로 화물역으로 폐쇄되어 동서를 분단하고 있던 지역이 도시의 회유성과 연계를 강화하는 거점적 퍼블릭 스페이스로 재탄생했다.

지역을 넘어 이어지는, 녹음이 연결되는 광장

북쪽의 프론트 포럼에는 벤치 등을 많이 배치하여 직장인의 휴식 공간 및 지역의 이벤트도 개최될 수 있는 공간으로 만들었다. 남쪽 가든 몰에는 녹색의 열대식물을 심어 아늑한 보행 공간을 형성하였고, 나무그늘에서 대화를 즐길 수 있게 벤치와 하이 카운터를 설치하였다. 건축주의 양해와 설계자 간의 자주적인 조정을 통해 건축 저층부는 차양 높이나 건물 셋백으로 통일된 지역 풍경을 만들어 사업공모에서 제안되었던 갤러리 등의 퍼블릭한 기능을 도입하여 직장인의 교류의 장을 만들었다.

배치 Site

가든 몰 단면 Section of Garden Mall

Focused public space with secured continuity with its surroundings thanks to high-level planning

상위 계획에 따라 주변과의 연속성을 담보로 한 거점적인 퍼블릭 스페이스

이 장에서는 민간 개발에 따라 정비되는 거점적인 퍼블릭 스페이스를 소개한다.

앞 장에서 다룬 보행자 네트워크뿐만 아니라 이러한 퍼블릭 스페이스의 대부분은 가이드라인과 지구 계획 등의 상위 계획에 따라 정해진 것들이다. 많은 경우가 민간 개발 부지와 주변의 공원 등이 일체적으로 조성되어 있다. 도쿄 미드타운에서는 옛 방위청 부지 개발 시 인접한 미나토구 히노키초 공원까지를 포함하는 범위로 지구 계획(오른쪽 그림)을 조성하여 히노키초 공원과 연속하는 형태의 공공 공지 정비를 계획함으로써 대규모 녹지 공간을 확보하고 있다(p102-105).

나카노역 북쪽 출구에 있는 경찰대학교 등의 부지 개발에서는 나카노역 주변 지역 만들기 가이드라인을 2007년에 제정하고, 중앙 도시 계획 공원(나카노 시키노모리 공원) 정비 및 그 주변의 민간 개발 부지에 공공 공지를 설치하도록 규정함과 동시에 나카노 4초메 지구의 지구 계획(오른쪽 아래)에도 반영시킴으로써 관민 일체의 거점적인 퍼블릭 스페이스를 실현하고 있다(p120-121).

더 중요한 것은 이들 주변 지역과의 연속성까지 배려하고 있다는 점이다. 위에 기술한 상위 계획 중에서도 공원 및 공공 공지 등을 계획하는 동시에 그 주변의 도로 등을 연결하는 보행자 네트워크를 지역 시설 등에 계획하여 주변 지역에서 접근하기 쉽고 연속된 퍼블릭 스페이스를 제공하고 있다.

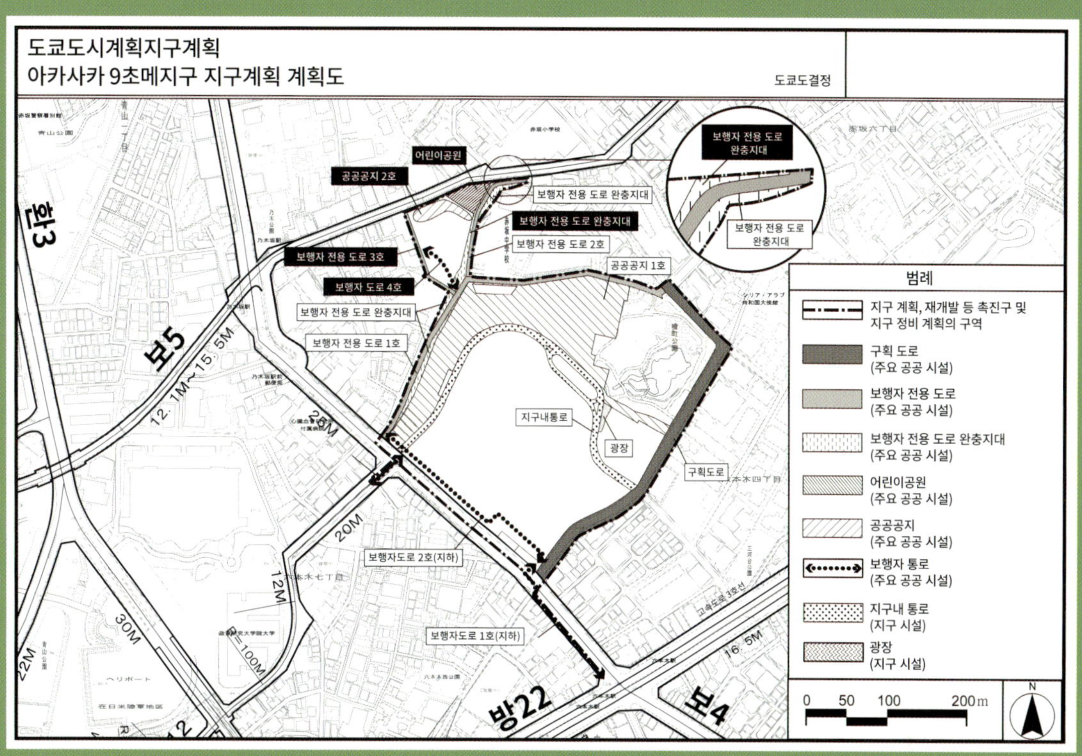

아카사카 9초메 지구의 지구 계획, 계획도
Plan diagram of Akasaka 9-chome District plan

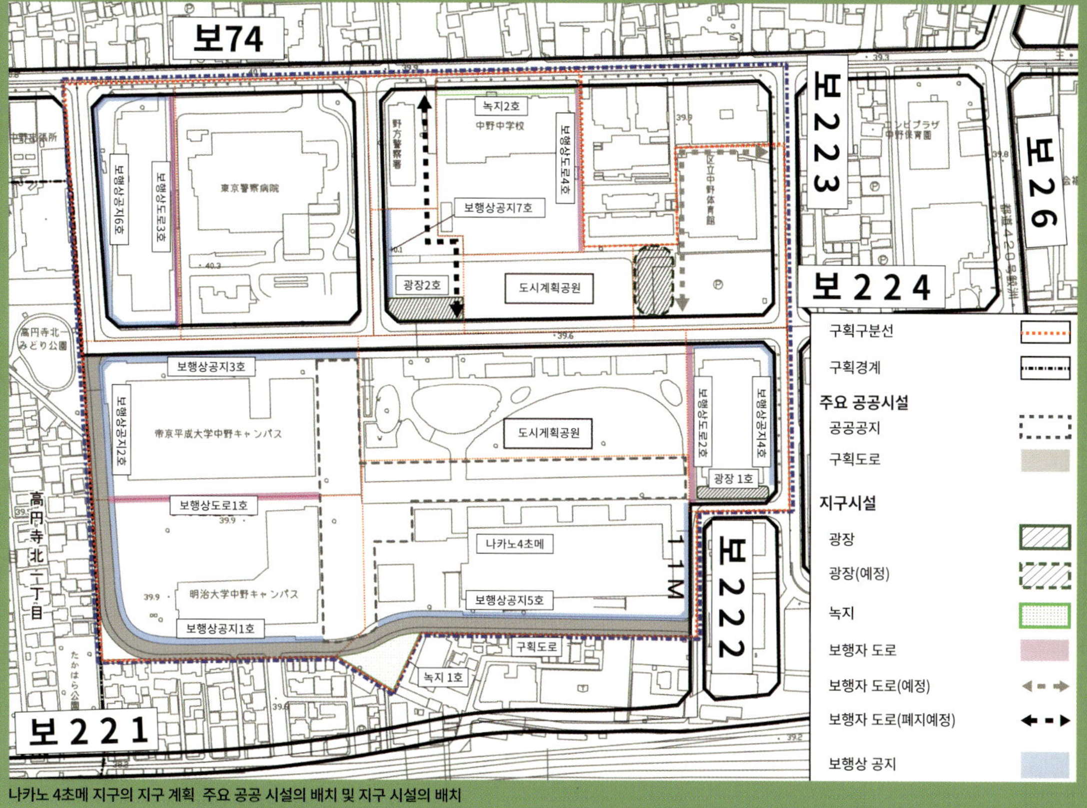

나카노 4초메 지구의 지구 계획 주요 공공 시설의 배치 및 지구 시설의 배치
Nakano 4-chome Area District Plan, Placement of Major Public Facilities and Placement of District Facilities

나카노 시키노모리 공원 (나카노구)
Nakano Shikinomori Park (Nakano City)

The bustle of privately owned public space seamlessly connecting with a park

—— Nakano Central Park　Park Avenue
Nakano Shiki no Mori Park

공공 공지의 활기가
공원까지 자연스럽게 연결

나카노 센트럴 파크　파크 에비뉴
나카노 시키노모리 공원(2012)

나카노역 북쪽 출구 지역의 약 16.8ha의 재개발에 있어서, 나카노 구립 나카노 시키노모리 공원 정비와 더불어, 공원과 원활한 활기를 창출하기 위해 민간 개발인 나카노 센트럴 파크의 공공 공지에 「파크 애비뉴」가 정비되었다. 이 공간은 카페의 테라스로 또는 비어 가든 등의 이벤트가 열릴 때 적극적으로 이용되어 공원·공공 공지가 연계된 일체적인 퍼블릭 스페이스가 탄생되었다.

120

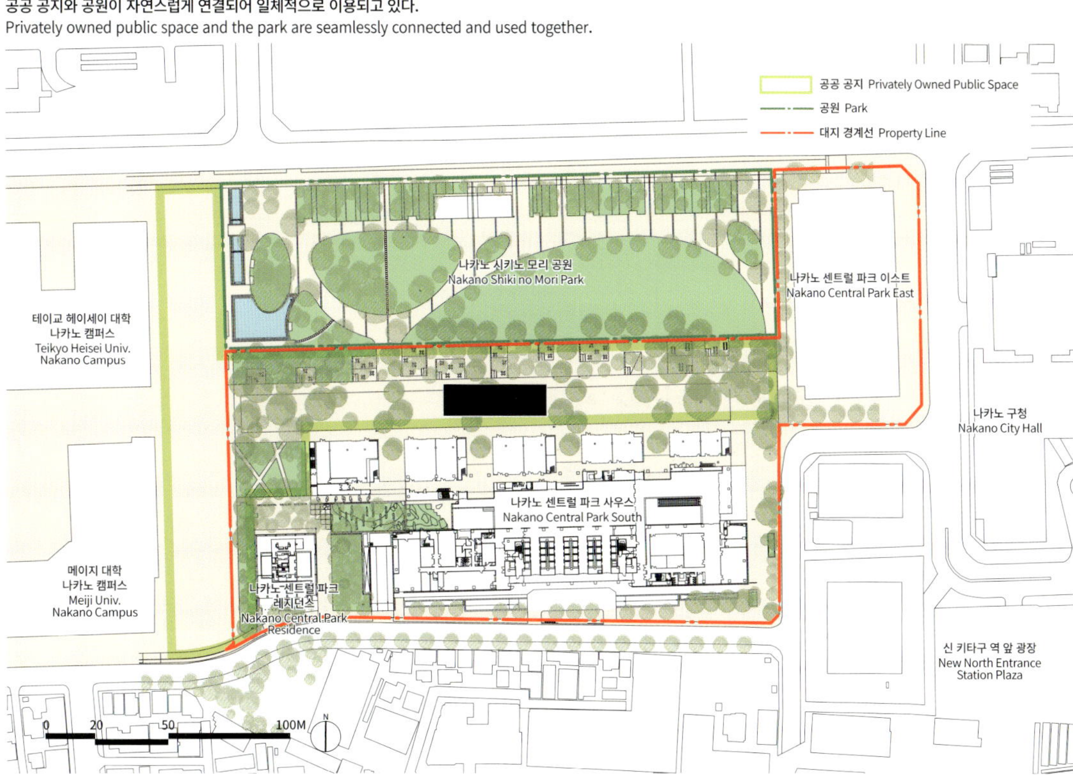

공공 공지와 공원이 자연스럽게 연결되어 일체적으로 이용되고 있다.
Privately owned public space and the park are seamlessly connected and used together.

배치　Site

Eco Corridor created with two city blocks

Keio Plaza Hotel' s Wood Promenade Garden
Shinjuku Fukutoshin Line 4　55HIROBA

2구획에 의한 Eco Corridor의 탄생

게이오 프라자 호텔 Wood Promenade Garden(1971)
신주쿠 부도심 4호선 55HIROBA(1974)

게이오 플라자 호텔(1971) 신주쿠 미츠이 빌딩(1974)

MAP p.195

니시 신주쿠 지역에서는 특정 지구 제도를 활용한 초고층 건축이 많이 입지해, 건물 저층부 공간에는 다양한 퍼블릭 스페이스가 탄생되고 있다. 그 중에서도 신주쿠 미츠이 빌딩과 게이오 플라자 호텔은 2구획이 인접한 부분에 무사시노 식물군을 재현한 녹지를 정비하고, 구획 도로 내의 식재도 포함하여 Eco Corridor를 형성하고 있다. 구획 단위의 개별 계획이고 특정 지구 제도의 틀을 넘어선 인접 지구 녹지와 호응하여, 니시 신주쿠 지역 개발 당시의 지역 만들기 콘셉트이기도 한 「살아있는 휴먼 스페이스의 재생」을 실현하고 있다.

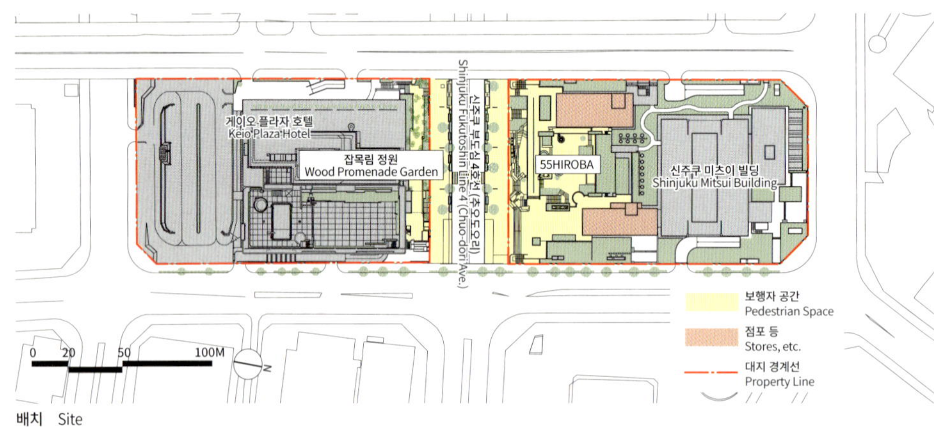

배치　Site

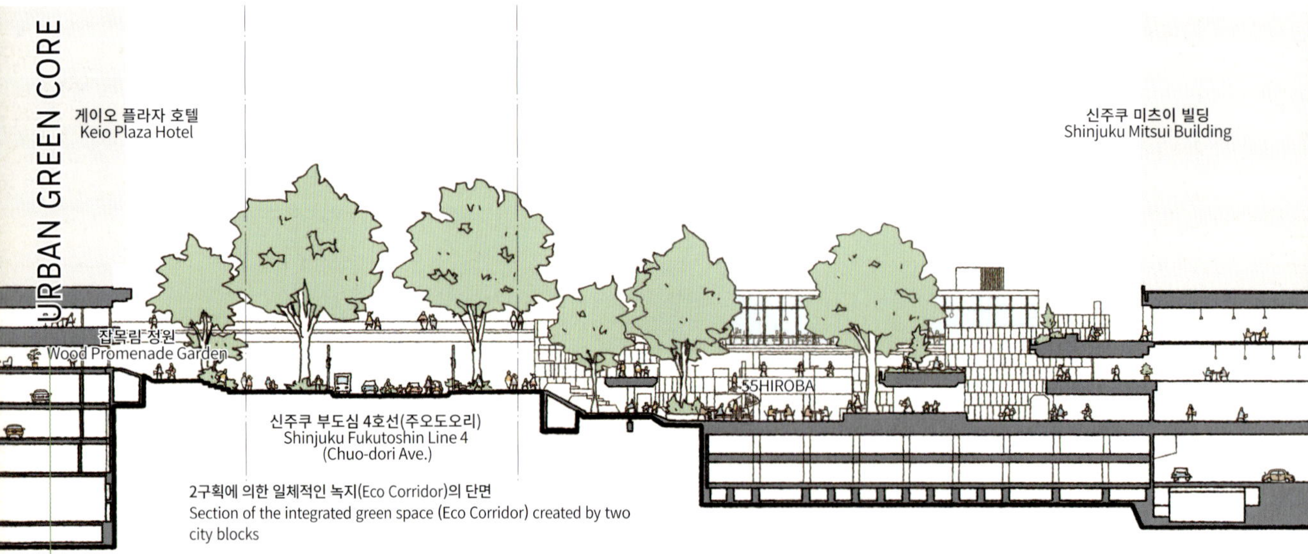

2구획에 의한 일체적인 녹지(Eco Corridor)의 단면
Section of the integrated green space (Eco Corridor) created by two city blocks

남서측에서 내려다 본 광경
Looking down from the southwest side.

신주쿠부도심 4호선(주오도오리)에서 55HIROBA를 바라봄
View of 55HIROBA from Shinjuku Fukutoshin Line 4 (Chuo-dori Ave.).

신주쿠도심 4호선(주오도오리)의 가로수와 일체된 진목린 정원
Wood Promenade Garden integrated with street trees.

가스미가세키 빌딩의 저층부에 설치된 약 1.3ha의 가스미가세키 테라스
Approx. 1.3ha Kasumi Terrace at the foot of Kasumigaseki Bldg.

블록 일체로 도시 공간의 리뉴얼

가스미가세키 테라스·케아키 광장(2009)

\가스미가세키 빌딩(1968) 가스미가세키 코몬 게이트(2007)

일본 최초의 초고층 빌딩인 가스미가세키 빌딩에서는 인접한 관청 건물의 재건축을 계기로 도시 계획 변경 및 저층부 계획 리뉴얼이 진행되어, 이 지역에서 일하는 사람들의 시대적 요구에 응한 도시 공간으로서의 재구축을 도모하고 있다.

Renewal of urban space through integration of city blocks

Kasumi Terrace, Keyaki Square

KASUMI DINING

BEFORE

남쪽에서 본 관경. 준공 당시의 광장
View from the south. The square at the time of completion.

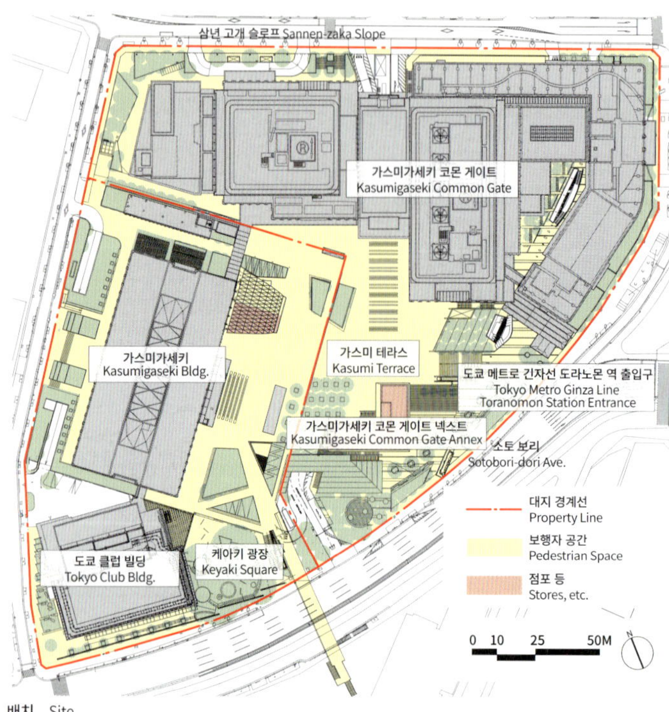

Expanding public space through revisions to city plans integrated with the adjacent city blocks

인접 지구와 일체적인 도시 계획 변경을 통한 퍼블릭 스페이스의 확충

가스미가세키 빌딩 건설 당시에 정한 특정 지구를 폐지하고 새롭게 인접 부지를 포함한 재개발 등 촉진 구역을 정하는 지구 계획을 구축하였다. 그 중에서 기존의 공개 공지를 포함한 일체적인 퍼블릭 스페이스를 정함으로써 광장의 기능 및 공간의 확충을 실현하였다. 이로 인해 약 1.3ha의 광장인 「가스미가세키 테라스」는 단구의 형태로 크게 열려 있고 지하철역에서의 동선도 개선하고 있다. 광장에는 녹지, 화단, 분수, 조명, 벤치, 아트 등을 배치하고, 가스미가세키 빌딩 준공 당시부터 크게 자란 느티나무의 보존 수목과 함께 계절감 있고 윤기 있는 풍경과 푸르름이 무성한 공간을 도심에서 창출하고 있다.

AFTER

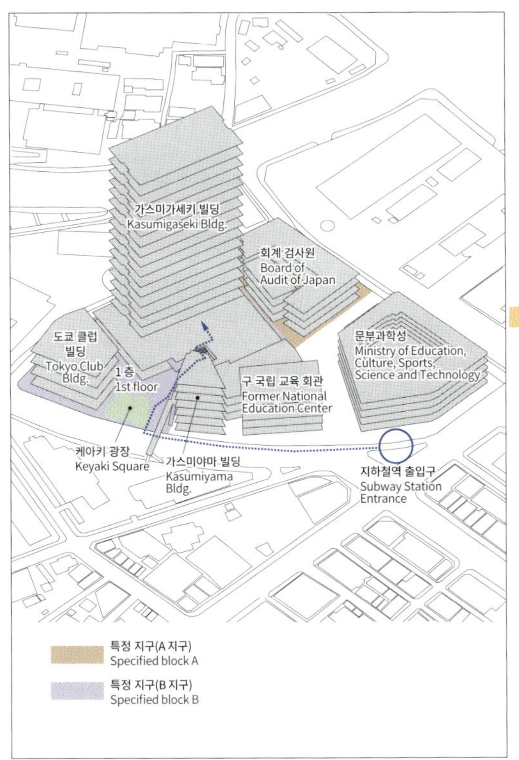

남쪽에서 본 관경. 리뉴얼 후 현재의 광장
View from the south. The current square after the renewal.

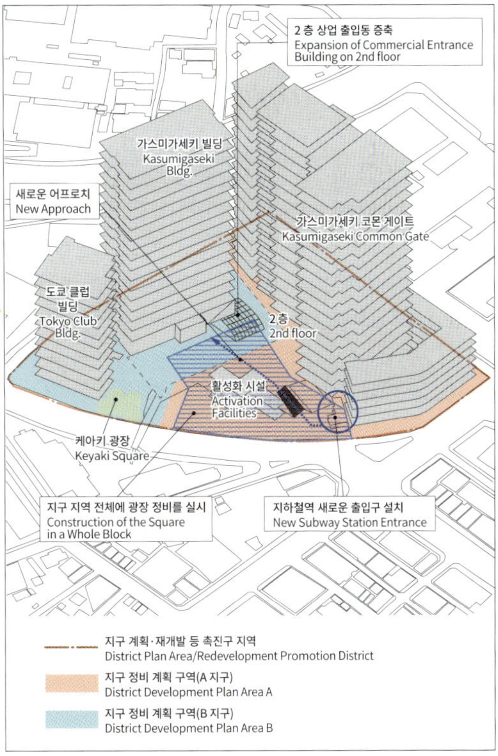

다이어그램　Diagram

Green Infrastructure - Revetment converted to park open space

Okawabata River City 21   Tsukuda Park, Ishikawajima Park

수변의 공원화로 인해 확장된 녹지

오가와바타 리버시티 21(2010) 츠쿠다 공원 이시카와시마 공원

스미다강을 따라 창고·공장 부지의 토지 이용 전환에 의해 창출된 오가와바타 리버시티에는 수변과 일체가 된 녹음이 풍부한 지역 만들기를 실현하고 있다. 수변 정비에 있어서 개발 당시에 일반적이었던 인공적인 형태의 제방을 대신하여 슈퍼 제방을 사용하여 상부를 공개 공원·녹지로 정비하고, 민지 내의 공개 공지와 일체적인 공간으로 만들 수 있도록 수변을 열려 있는 거점적인 퍼블릭 스페이스를 창출하여 사람들이 산책이나 조깅 코스, 쉼터 등으로 다양하게 이용할 수 있게 하였다. 이 수변 만들기 및 공간 만들기 기법은 이후 강변의 지역 만들기로도 이어지고 있다.

이시카와시마 공원의 남쪽에서 본 관경
View from the south side of Ishikawajima Park.

주오오하시에서 츠쿠다 공원의 북쪽을 본 관경
View of the north of Tsukuda Park from Chuo-ohashi Bridge.

배치　Site

단면　Sectoin

MAP p.194

35

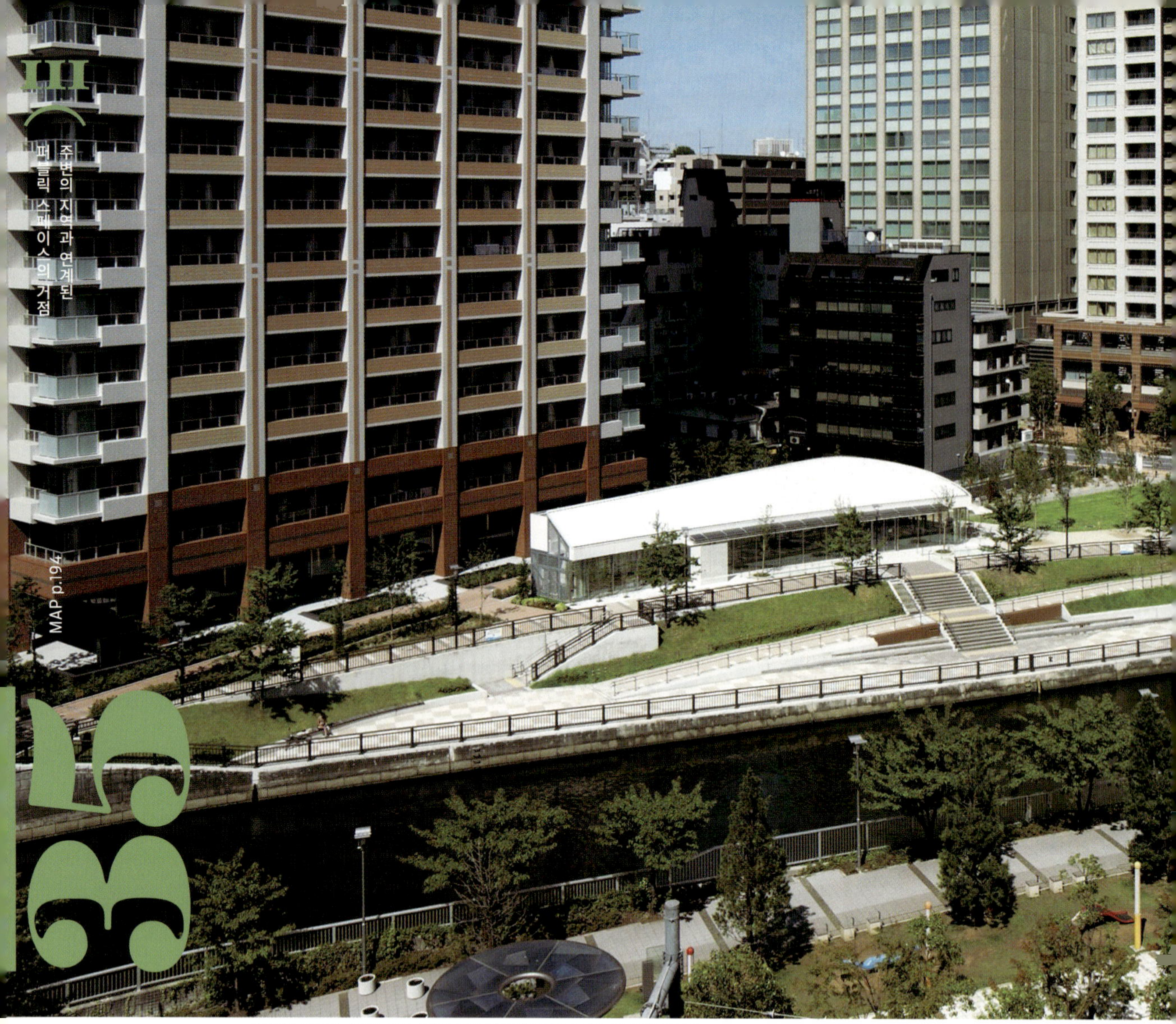

서쪽에서 본 관경. 공원화, 방파제의 일체 정비에 의해 실현한 고탄다 수변 만남의 광장
View from the west. Gotanda Fureai Waterfront Plaza turned into a park and integrally developed with the revetment.

배치 Site

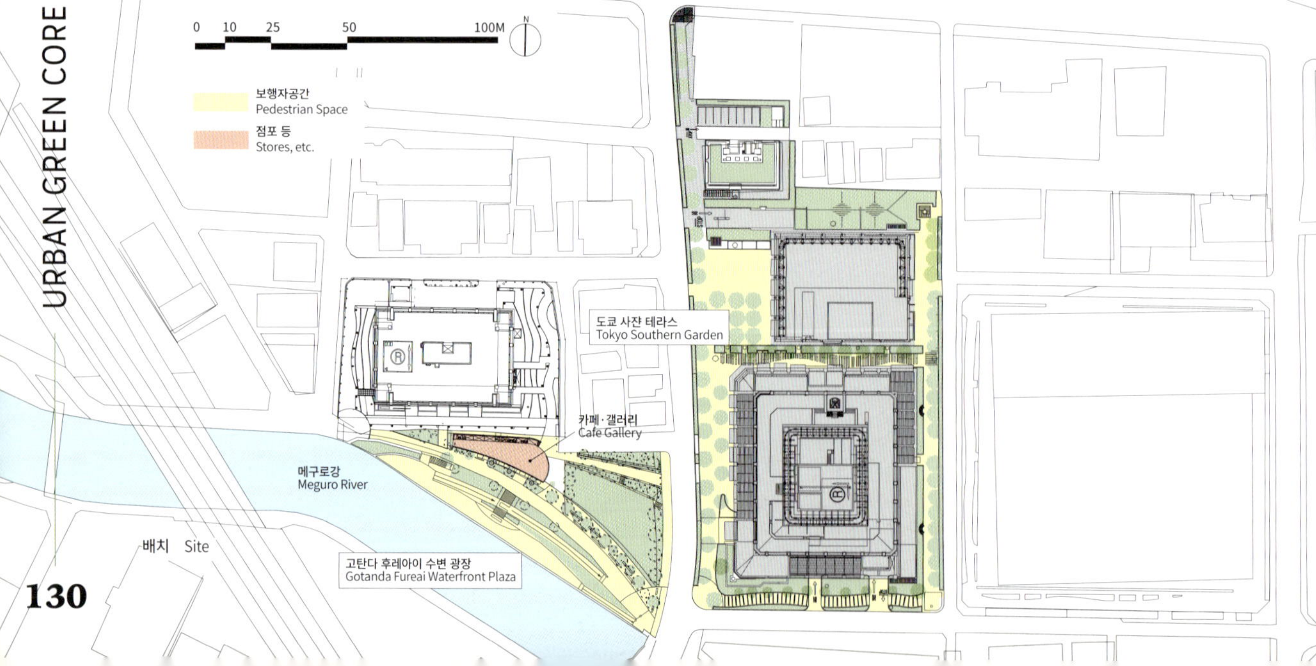

Creating hydrophilic space that doubles as revetment by making the site a park

Tokyo Southern Garden   Gotanda Fureai Waterfront Plaza

부지의 공원화로 인해 방파제를 겸한 친수 공간을 형성

도쿄 사잔 가든 고탄다 만남의 수변 광장(2010)

메구로강을 따라 정비된 「고탄다 수변 만남의 광장」은 지역과 강을 잇는 거점적인 도시 공간으로서 사계절의 다양한 표정을 보이며 사람들에게 사랑받고 있다.

종전의 메구로 강변은 전형적인 방파제로 일률적인 정비가 이루어져 있고 지역과 강이 떨어져 있었지만, 강변 개발을 계기로 택지의 일부를 공원화 및 친수 방파제화하여 하천 구역으로 지정함으로써 가깝게 강을 느낄 수 있는 친수 공간을 창출하였다.

과거 메구로 강변의 모습
View along Meguro River before.

현재 친수 공간의 모습
View of the current hydrophilic space.

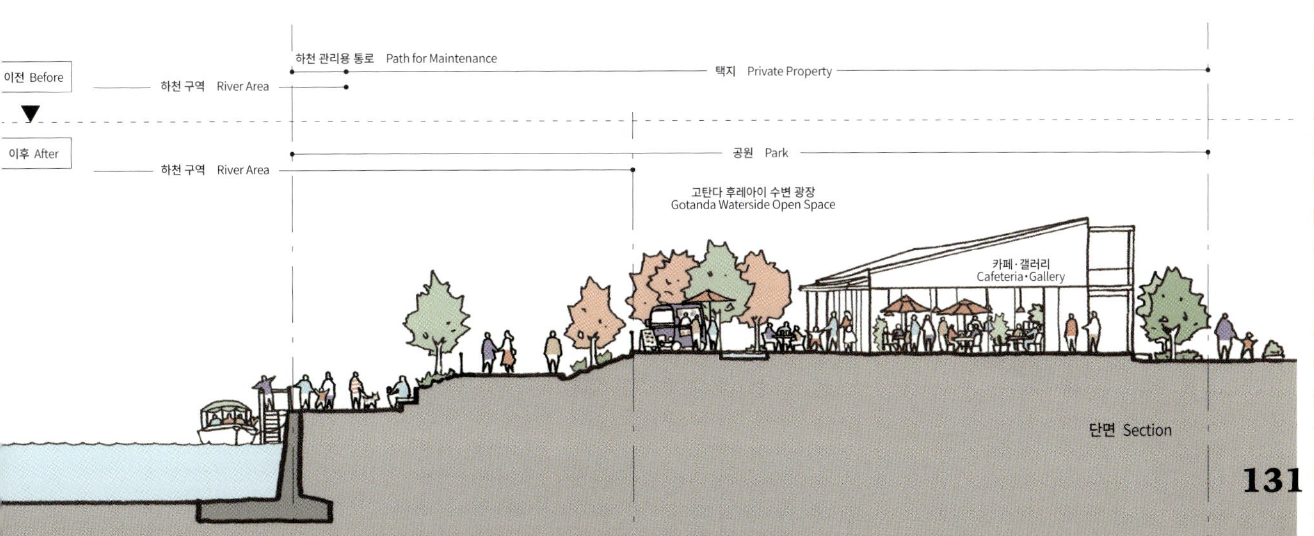

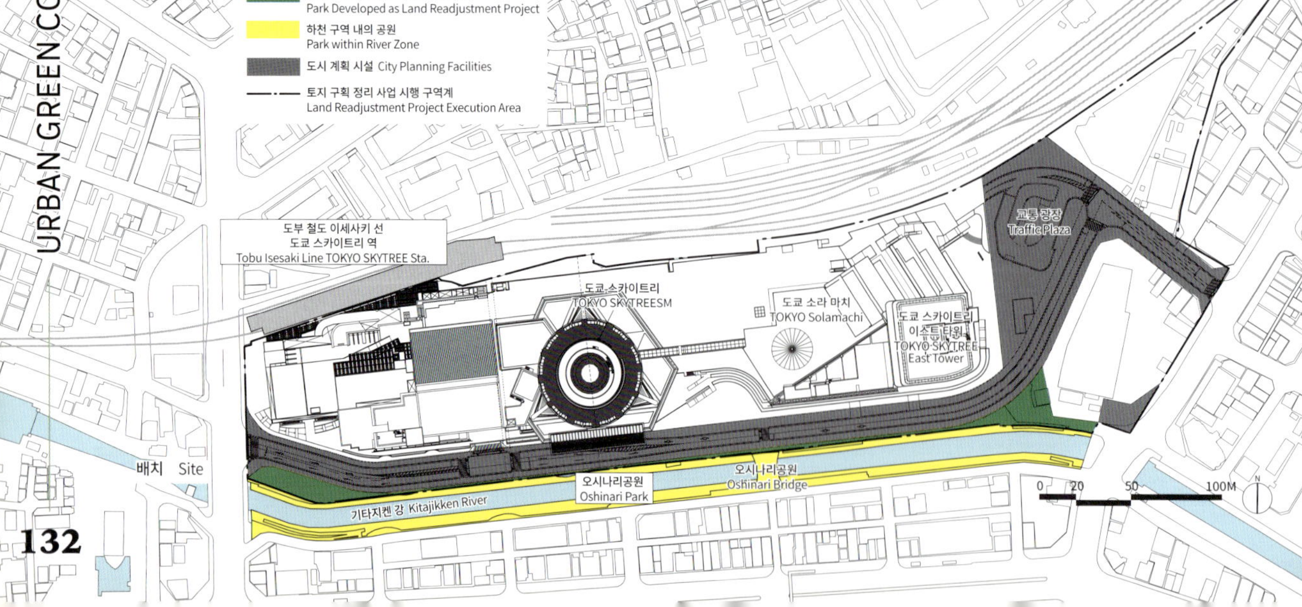
URBAN GREEN CORE

토지 구획 정리 사업으로 정비된 공원
Park Developed as Land Readjustment Project
하천 구역 내의 공원
Park within River Zone
도시 계획 시설 City Planning Facilities
토지 구획 정리 사업 시행 구역계
Land Readjustment Project Execution Area

도부 철도 이세사키 선
도쿄 스카이트리 역
Tobu Isesaki Line TOKYO SKYTREE Sta.

도쿄 스카이트리
TOKYO SKYTREESM
도쿄 소라 마치
TOKYO Solamachi
도쿄 스카이트리
이스트 타워
TOKYO SKYTREE
East Tower
교통 광장
Traffic Plaza

오시나리공원
Oshinari Park
오시나리공원
Oshinari Bridge
기타지켄 강 Kitajikken River

배치 Site

0 20 50 100M

Development of TOKYO SKYTREE TOWN, creation of integrated waterfront space

— TOKYO SKYTREE TOWN Oshinari Park

도쿄 스카이트리 타운® 개발과 일체 정비된 수변 공간

도쿄 스카이트리 타운® 스미다 구립 오시나리 공원(2012)

도쿄 스카이트리 타운의 개발과 함께 인접한 공공 공간에서도 공원 정비 및 방파제 정비 등이 일체적으로 이루어지고 있다. 하천 구역과 인접한 부분에서는 토지구획정리사업을 통해 기부 공원이 정비되고, 하천 구역 내에서는 도쿄도가 방파제 정비(내진화·호안컷·프롬나드 설치)를 한 후 스미다구가 수경 공사 등의 친수 공간 정비와 인도교 정비를 실시하여, 이들을 통칭하여 「스미다구립 오시나리 공원」으로 구성하였다. 민관의 여러 정비 주체가 일체의 연속된 공간을 창출하고 있는 것이 특징이며, 수변을 이용하는 사람들의 액티비티를 풍부하게 창출하는 친수 공간이 탄생하고 있다.

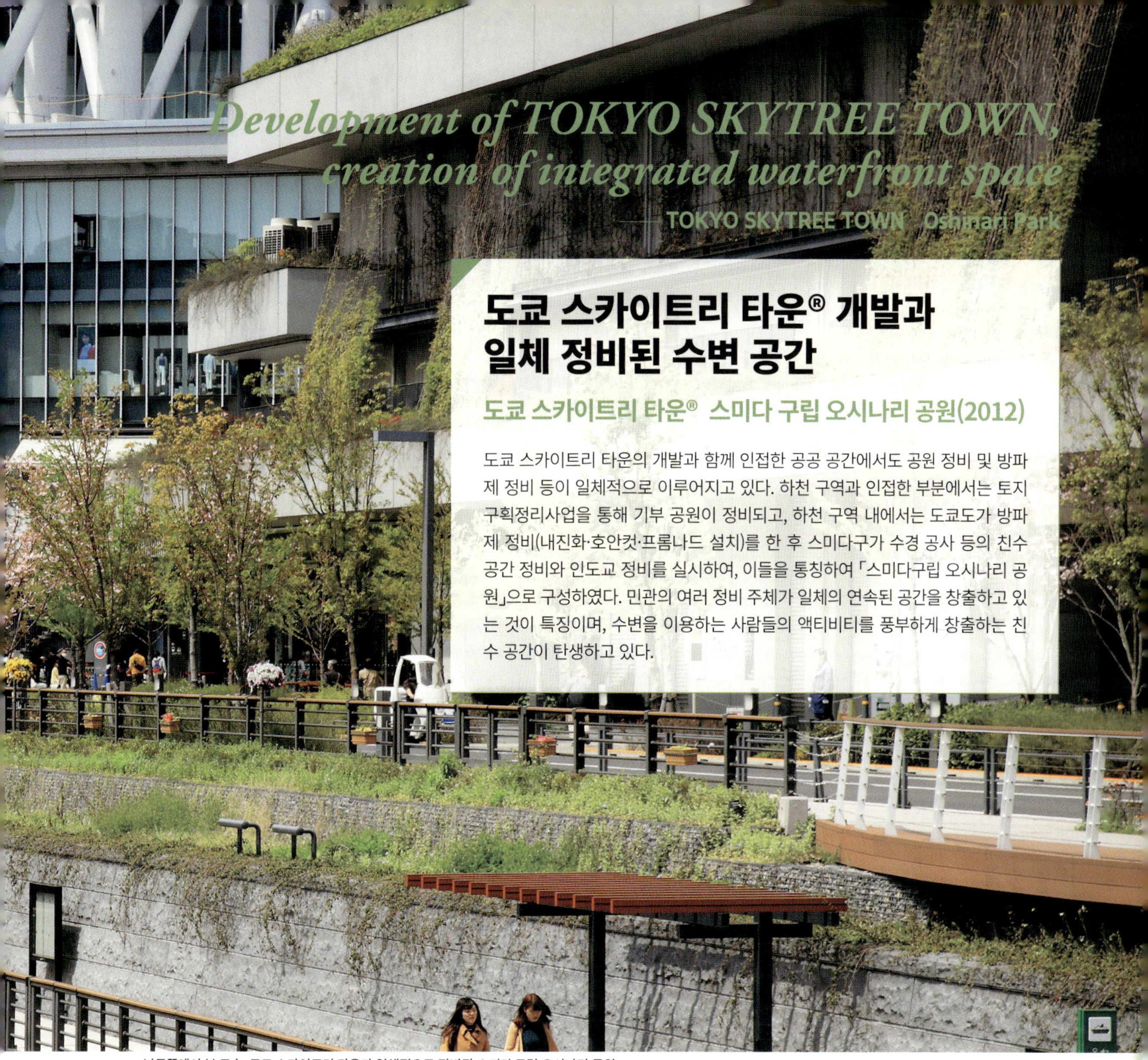

남동쪽에서 본 모습. 도쿄 스카이트리 타운과 일체적으로 정비된 스미다 구립 오시나리 공원
View from the southeast. Oshinari Park in Sumida City was developed integrally with TOKYO SKYTREE TOWN.

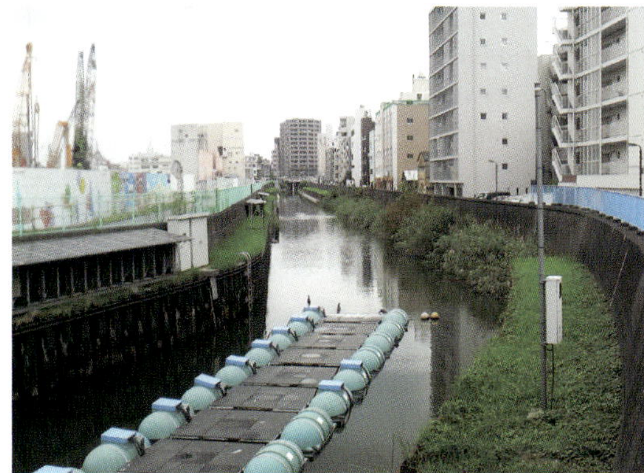

이전의 키타지켄강
Kitajikken River before.

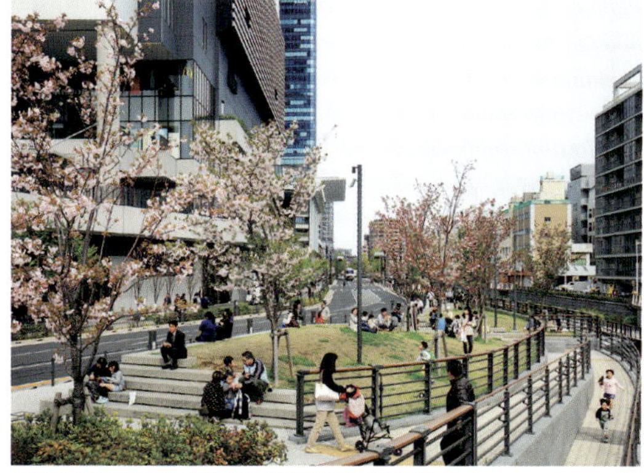

이후의 키타지켄강에는 하천 구역 내외가 일체 정비되었다.
After redevelopment, both public and private land inside and outside the Kitajikken river boundary has been integrally improved.

Realization of collaborative public space through planning guidance

Kita-aoyama 3-chome District Project

ANTICIPATED PROJECT

계획 유도에 의한 협력형 퍼블릭 스페이스의 실현

키타아오야마 3초메 지구 지역 만들기 프로젝트

도영 아오야마기타쵸 아파트의 재건축을 계기로 한 키타아오야마 3초메 지역 만들기에서는 재건축·갱신하는 건물과 기성 시가지 사이에 대규모의 광장이 탄생된다. 미나토구가 작성한 「아오야마도오리 주변 지구 지역 만들기 가이드라인(2015년 10월)」과 도쿄도가 만든 「사업 실시 방침(2016년 1월)」에서 지역 만들기의 방향성이 제시된 가운데, 3가지 다른 사업으로도 단계적으로 재건축·갱신이 진행되는 상황에서도 관계자 간 협력을 도모하여 일체적인 지역 만들기의 실현을 위해 프로젝트를 진행하고 있다. 그 중 두 개의 선행 사업에 의한 성과 중 하나가 이 광장이다.

선행 사업으로 정비된 녹색 광장은 향후 예정되어 있는 재개발 사업을 통해 더욱 대규모로 확장될 예정이며, 녹음이 풍부한 활력·교류의 거점으로써 지역의 휴식처가 탄생될 것이다.

3가지의 사업에 따른 미래 정비 이미지
Image of future development by the three projects.

광장의 산책로. 재건축을 통해 창출된 풍부한 녹지는 향후 예정되어 있는 재개발 사업을 통해 더욱 확장될 것이다.
A promenade in the plaza. The affluent greenery created by the rebuilding will be further expanded by the planned redevelopment project.

부지 내 광장에서 본 관경
View from the plaza on the property.

부지 내의 비오톱
Biotope on the property.

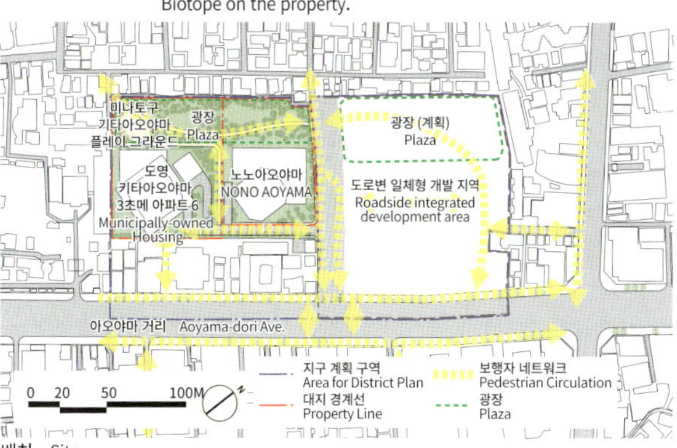

「아오야마도오리 주변 지구 지역 만들기 가이드라인」에 있는 지역 만들기의 방향성
The image of the initiatives in Aoyama Street Community Guideline for Community Development

배치　Site

135

BARGAIN!
2020.10.01

IV

MULTI-LAYERED DEVELOPMENT

Conglomerated public spaces crossing
different domains

**영역 횡단형
복합 퍼블릭 스페이스**

도로 상공의 오픈 공간

토라노몬힐즈 오발(Oval) 광장(2014)

토라노몬힐즈는 입체 도로 제도의 활용에 의해 도시계획 도로 환상 2호선을 건축의 지하 부분에 내포해 도로와 시설을 일체로 정비한 복합 개발이다. 이 도로 상부에는 녹음이 풍부한 퍼블릭 스페이스인 「오발 광장」이 만들어졌다. 특징적인 큰 지붕 아래 예술과 자연을 즐기는 일상의 풍경을 만듦과 동시에 다양한 이벤트가 열리는 도시의 오픈 광장이 되고 있다.

Festive space above the road

Toranomon Hills   Oval Plaza

다양한 활동과 자연을 즐길 수 있는 공간을 실현한 토라노몬힐즈 오발 광장
Toranomon Hills Oval Plaza is a space for enjoying various activities and greenery.

IV
복합 퍼블릭 스페이스
영역 횡단형
38
MULTI-LAYERED DEVELOPMENT
입체도로제도를 활용하여 도로위 광장을 정비했다.
Multi-level road system was leveraged to create the plaza above the road.
TORANOMON HILLS
민간부지
Private Property
도로구획(입체도로)
Road Area(Multi-level Road)

광장의 경사를 활용한 액티비티의 창출
Conducting activities that utilize the slope of the plaza.

Activities utilizing sloped plaza

광장의 경사를 활용한 액티비티

환상 2호선이 부지 내 지하에서 지상으로 올라오기 때문에 도로 상공에 설치된 오발 광장에도 경사가 있다. 이로 인해 광장에는 이 경사를 활용하여 무대와 관객석을 만들었다. 또한 도로 상공의 특성을 살려 광장 단부에서 도로를 바로 아래로 내려다볼 수 있는 공간을 만드는 등, 도시의 새로운 액티비티를 유발시켰다.

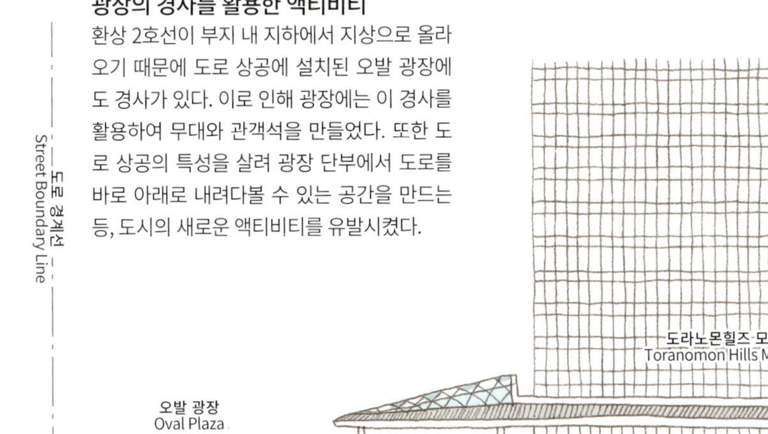

단면 Section

역과 바로 연결된 도로 위의 광장

롯본기 힐즈 66 플라자 Metro Hat(2003)

롯본기 힐즈 모리타워의 저층부, 환상 3호선 상공의 인공 지반 위에 약 4,000㎡의 대규모 보행자 공간인 66 플라자가 조성되었다. 지하철 롯본기역에서 수직동선인 Metro Hat의 에스컬레이터를 오르면, 녹지와 수경, 모리타워가 눈앞에 펼쳐져 롯본기 힐즈의 현관으로서 많은 사람들이 일상적으로 오고 가는 만남과 휴식을 위한 장소를 제공하고 있다. 또한 66 플라자 하단에는 연결 측도와 부지 내 차로의 출입구가 있고 그 아래에는 아자부 터널이 있는 3층 구조로 되어 있어, 광역 교통망의 향상을 도모할 뿐만 아니라 니시아자부로 연결되어 지역의 연속성을 실현하고 있다.

환상 3호선에 정비된 66 플라자에서는 다양한 이벤트가 개최되어 사람들로 붐비는 모습을 보인다.
66 Plaza built above the Loop Road No. 3 is filled with people during various events.

북동쪽에서 66 플라자를 내려다본 광경
Looking down at 66 Plaza from the northeast.

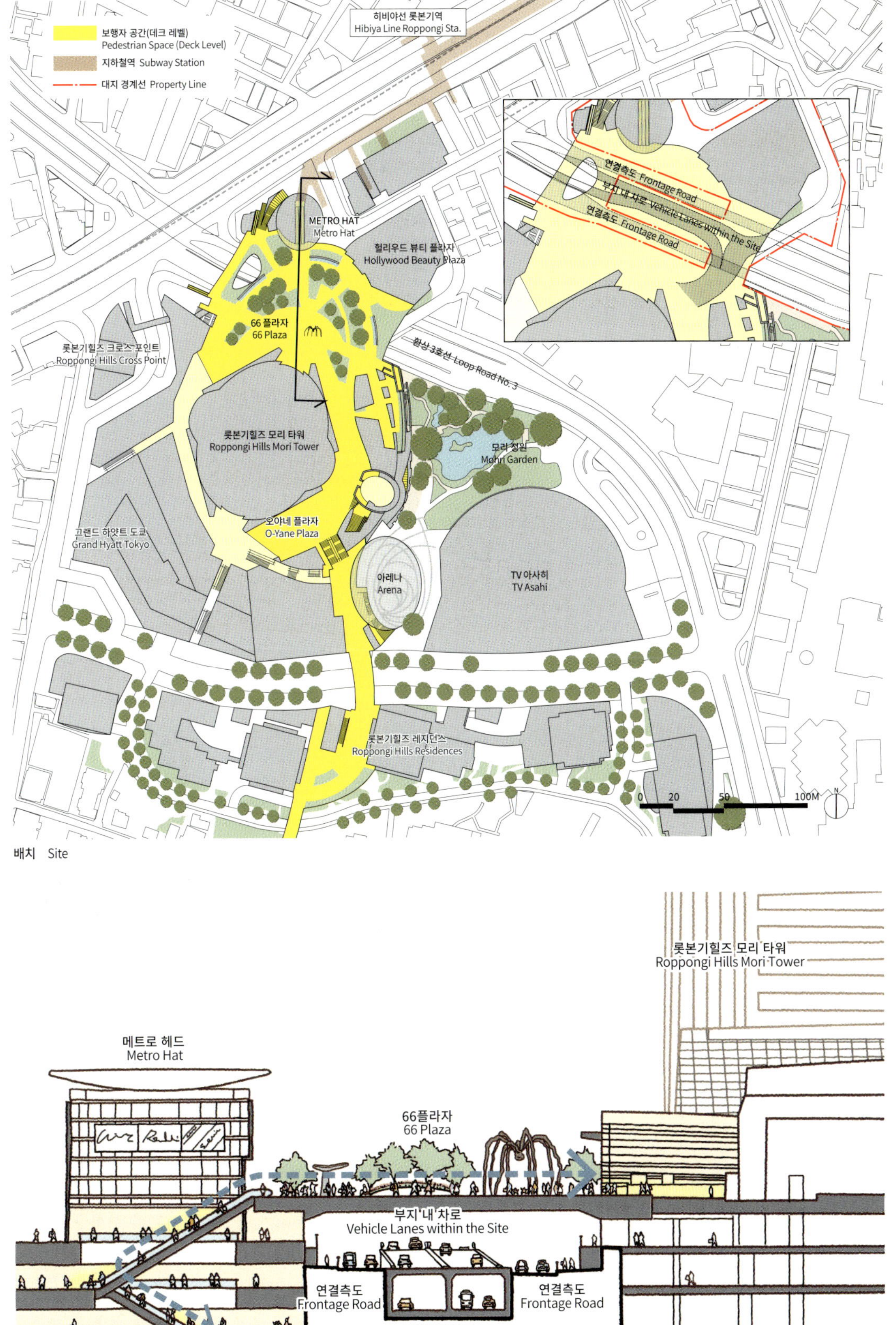

배치 Site

단면 Section

Open space above a sewage treatment plant

Shinagawa Season Terrace Event Square
Shibaura Chuo Park

하수도 인프라 위의 광장

시나가와 시즌 테라스 이벤트 광장
시바우라 중앙공원(2015)

하수처리시설(도시시설)에 입체 도시 계획 제도를 적용하고 인공지반 상부를 건축 부지로 조성하여 고층타워 및 퍼블릭 스페이스를 창출했다. 도쿄도가 소유하는 건축 부지 부분에 민간사업자의 정기 임차지권을 설정하고, 임차권의 대가로 하수 인프라 갱신 및 민간 개발 사업의 양립을 실현하고 있다. 인접한 미나토 구립 시바우라 중앙공원과 일체적으로 조성되어 넓은 광장 공간을 도시에 제공하고 있다.

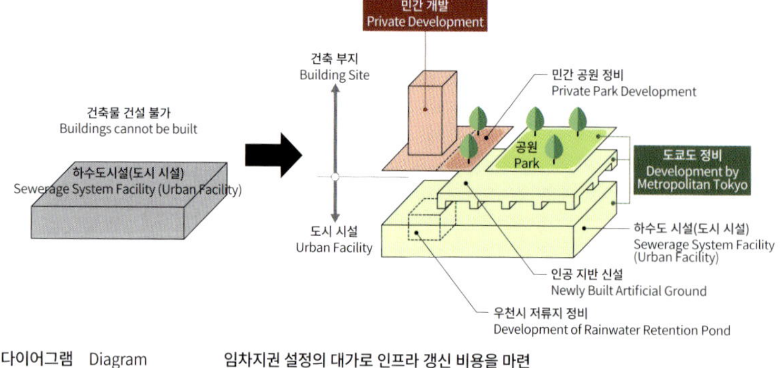

다이어그램   Diagram

임차지권 설정의 대가로 인프라 갱신 비용을 마련
Cost for infrastructure renewal was raised in exchange for the establishment of land lease rights.

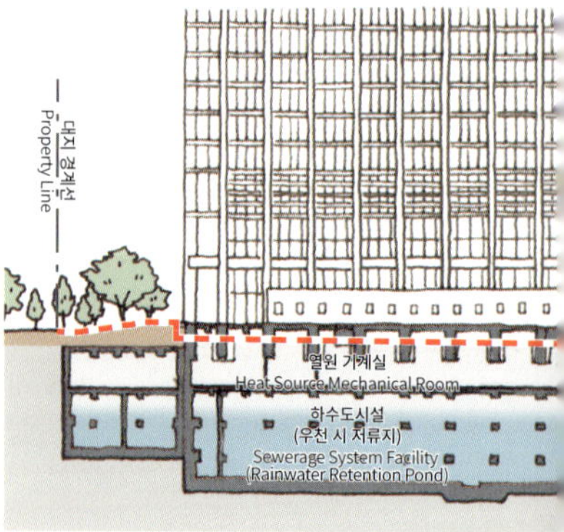

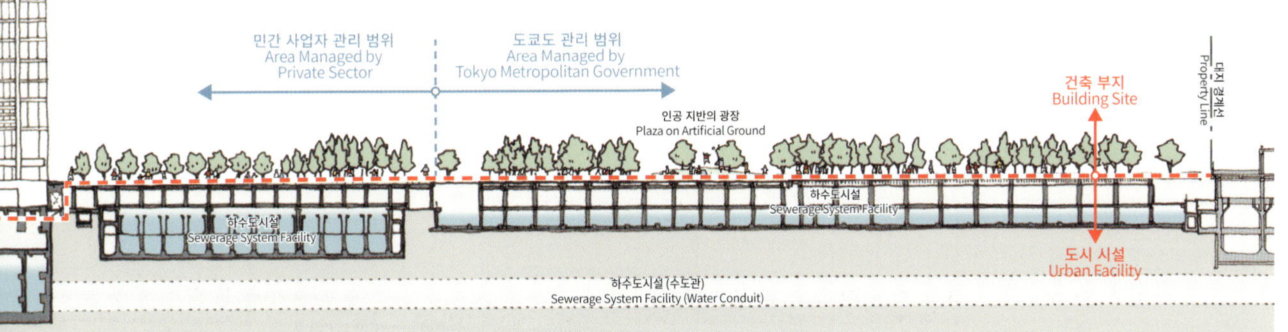

하수도 시설 상부에 만들어진 시나가와 시즌 테라스와 시바우라 중앙공원
Shinagawa Season Terrace and Shibaura Chuo Park built above the sewage treatment plant.

단면　Section　도시계획으로 하수시설의 입체적 구획을 하고 인공지반 상부를 건축 부지화함
Multi-level area of sewage facility was determined by city planning, and the areas above the
artificial ground were turned into building sites.

Development methods by type
for cross-domain public space

영역 횡단형 공공 공간의 유형별 정비 수법

RIVER × PUBLIC SPACE
하천×퍼블릭 스페이스

하천 점용은 하천 관리자의 승인이 필요하다 (하천법 24조). 허가 기준은 하천 점용 준칙으로 정한다.

기존의 준칙에는 점용 허가 대상으로 오픈 카페 등의 민간사업자가 포함되지 않았지만, 수변공간의 적극적인 활용을 목적으로 일정한 조건하에 민간사업자의 이용이 허용되었다.

① 2011년도 개정: 민간사업자의 영업 활동이 일부 허용

② 2013년도 개정: 민간사업자의 점용 허가 기간이 3년에서 10년으로 연장

하천 점용의 이미지
Image of occupying river.

ROAD × PUBLIC SPACE
도로×퍼블릭 스페이스

도로 공간 내부를 입체적으로 활용해 정비하는 퍼블릭 스페이스는 도로법에 의해 규정된 「도로 구조물」, 「겸용 공작물」, 「상공도로 등 (도로 점용)」이 일반적이다. 여기에 더해 도로법, 도시계획법, 건축기준법을 횡단적으로 활용한 「입체도로제도」를 활용하여 도로 공간 내에 민간 시설을 계획하는 것도 가능해진다.

입체도로제도의 이미지.
Image of multi-level road system.

소유자가 다른 공간의 영역인 횡단형 공공 공간 실현을 위해 다양한 관계 법령과 제도를 활용한 방법으로 지금도 시대의 요구에 맞춰 개량·신규 수법의 개발이 진행되고 있다. 퍼블릭 스페이스의 계획 단계에서는 실현되어야 할 공간의 특성에 맞는 가능한 방법을 검토하고 선택해야 한다. 여기서는 국내 주요 도시 개발 사례에서 활용된 방법을 중심으로 소개한다.

ARCHITECTURE × PUBLIC SPACE

건축×퍼블릭 스페이스

도시재생에 대한 기여 정도에 따라 용적률 등의 완화를 인정하는 도시재생 특별지구 적용으로 사업자가 창의적인 노력을 기울인 다양한 공간이 공헌 요소로서 인정받게 되었다. 예를 들어, 컨퍼런스 센터나 극장 등의 공공적 성격이 높은 시설과 일체적으로 이용 가능한 광장 공간이나 건물을 함께 입체화하는 것이 입지 특성상 유효하게 여겨지는 공지에 대해서는 건물 위라 하더라도 공공적 공간으로 인정받고 있다. 이러한 배경으로 건축과 공공적 공간이 융합된 새로운 퍼블릭 스페이스가 창출되고 있다.

건축물 상부의 퍼블릭 스페이스 이미지
Image of public space above building.

입체도시공원제도의 이미지
Image of multi-level urban park system.

도시 공원법에 따른 「입체도시공원제도」를 활용하여 공원의 입체적인 범위를 결정함으로써 공원 하부에 공원 이외의 기능을 설치할 수 있게 되었다. 원래 도시 공원에 설치할 수 있는 시설은 용도, 규모에 큰 제한이 있었지만, 본 제도를 활용함으로써 지역 특성에 맞는 보다 자유도가 높은 공원 계획이 가능하게 되었다.

광장 아래쪽의 시부야 강
Shibuya River Under the Plaza

시부야역 C2 출구에서 이나리 다리 광장, 시부야 스트림을 바라본 관경. 시부야강 위의 점용 부분과 민간 부지의 공간에 형성된 일체적인 퍼블릭 스페이스
View toward Inari Bridge Square and SHIBUYA STREAM from C2 exit of Shibuya Station. Integrated public space stretches between occupied space above Shibuya River and private property.

관민 협력을 통해 재생된 시부야강과 강변의 시부야 리버 스트리트   Shibuya River, revitalized through a public-private partnership, and Shibuya River Street along the river.

이벤트 시의 이나리 다리 광장
Inari Bridge Square during an event.

Bustling plaza created
above the river

Inari Bridge Square, Konno Bridge Square
in front of SHIBUYA STREAM

강 위에 만들어진 활기찬 광장

시부야 스트림 앞 이나리 다리 광장·
킨노 다리 광장(2018)

시부야 스트림의 도시 계획에 따른 공공 기여로서 수변의 활기 조성을 추진하는 목적으로 하천 상공 점용에 의한 광장 공간이 정비되었다. 정기적으로 마르세나 음악 라이브, 지역 축제 등의 이벤트가 개최되어 활기를 느낄 수 있는 건축 부지 내외가 일체적으로 연속하는 퍼블릭 스페이스가 생겨났다.

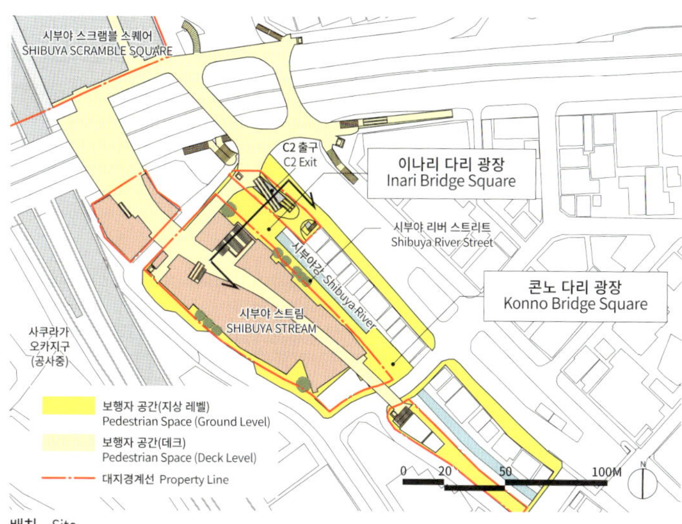

배치   Site

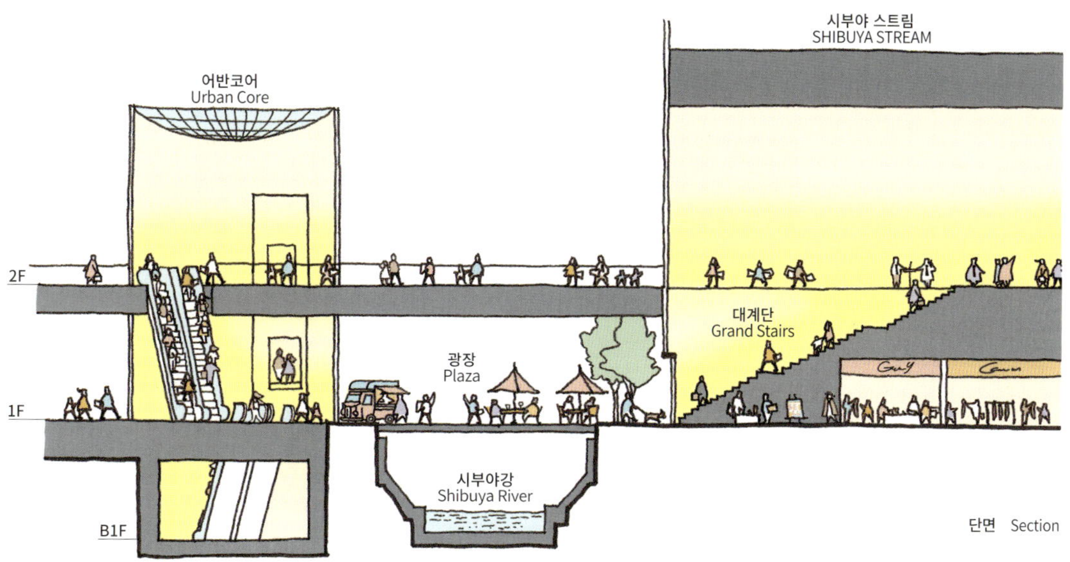

단면   Section

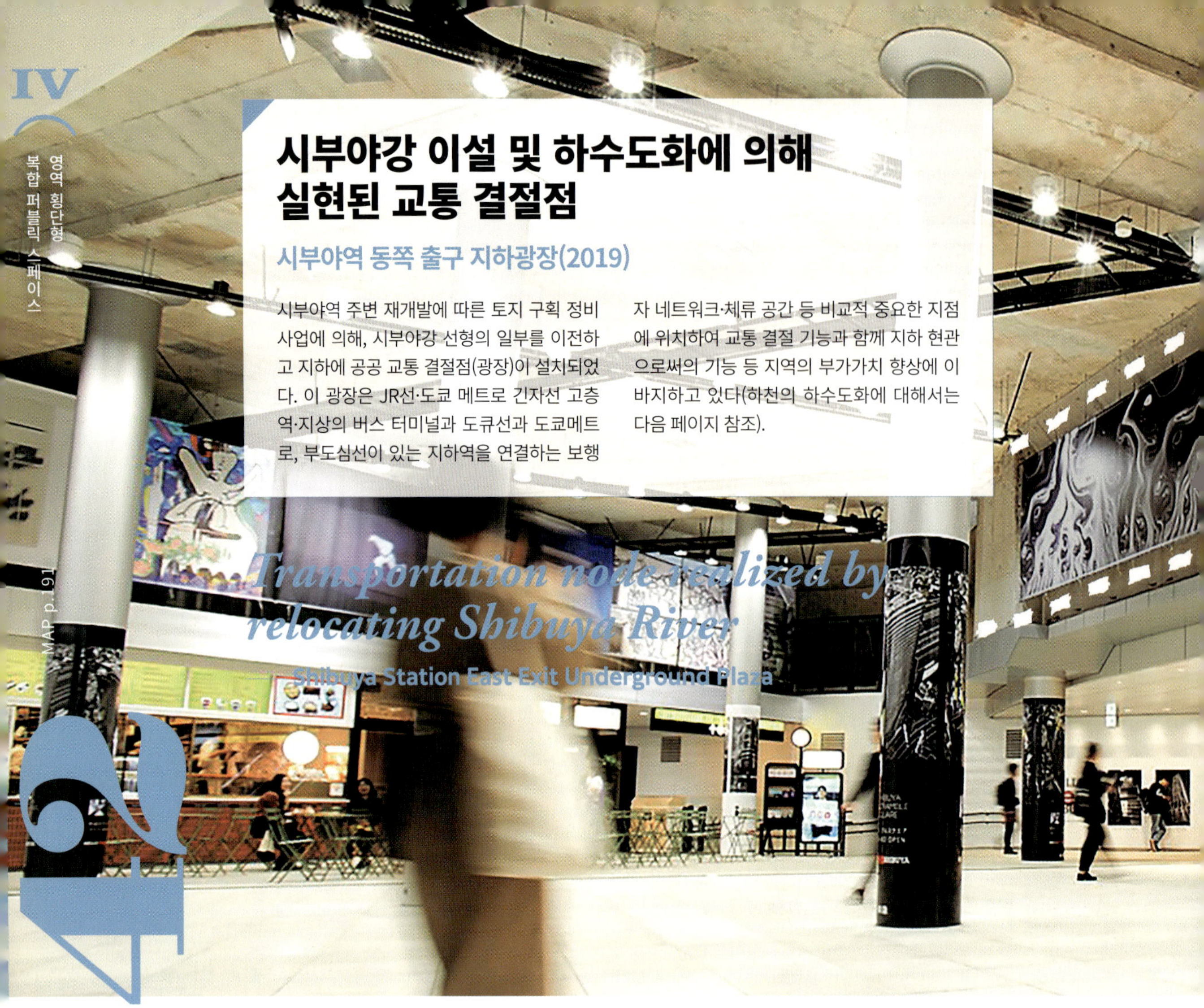

시부야강 이설 및 하수도화에 의해 실현된 교통 결절점

시부야역 동쪽 출구 지하광장(2019)

시부야역 주변 재개발에 따른 토지 구획 정비 사업에 의해, 시부야강 선형의 일부를 이전하고 지하에 공공 교통 결절점(광장)이 설치되었다. 이 광장은 JR선·도쿄 메트로 긴자선 고층 역·지상의 버스 터미널과 도큐선과 도쿄메트로, 부도심선이 있는 지하역을 연결하는 보행자 네트워크·체류 공간 등 비교적 중요한 지점에 위치하여 교통 결절 기능과 함께 지하 현관으로써의 기능 등 지역의 부가가치 향상에 이바지하고 있다(하천의 하수도화에 대해서는 다음 페이지 참조).

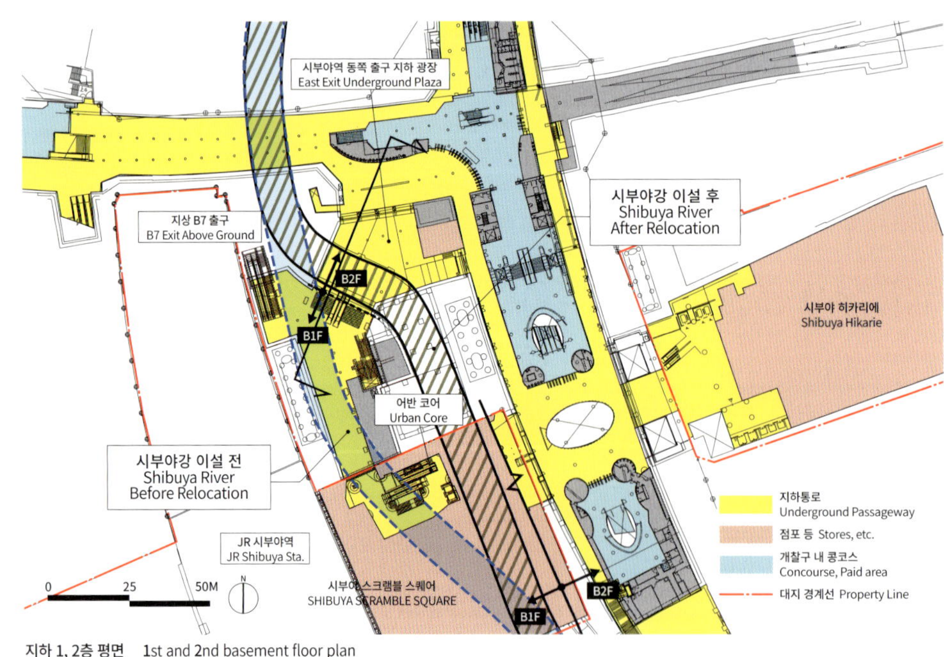

지하 1, 2층 평면   1st and 2nd basement floor plan

Transportation node realized by relocating Shibuya River
— Shibuya Station East Exit Underground Plaza

시부야강 아래를 통과하는 시부야역 동쪽 출구 지하광장
Shibuya Station East Exit Underground Plaza passing under Shibuya River.

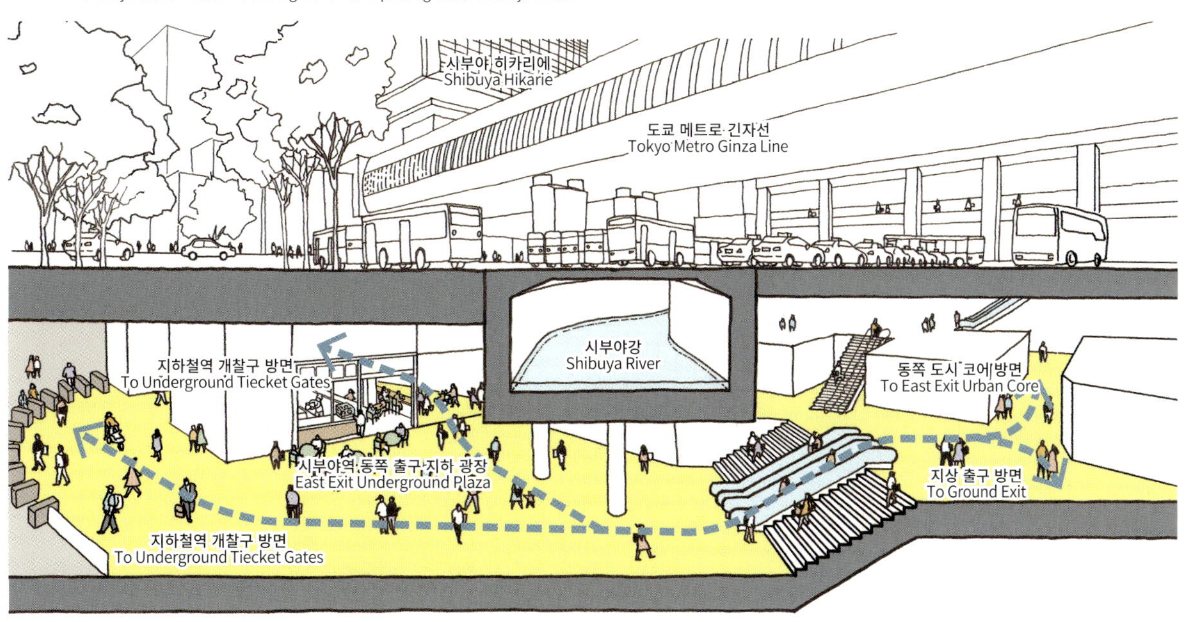

단면   Section

Shibuya River revitalization through collaboration between urban infrastructure development project and private development project

도시기반 정비사업과 민간개발사업의 협업에 의한 시부야강 재생

앞의 두 사례는 동일한 시부야강 위와 아래에 만들어진 퍼블릭 스페이스다. 모두 동선과 체류 공간의 확보·환경 개선·방재 기능 등 다양한 각도에서의 검토를 근거로 개별도시 개발사업뿐만 아니라 도시 기반 정비사업을 연계하여 하천을 따라 여러 개의 프로젝트와 여러 주체가 협력하여 실현하고 있다.

구획 정리 사업 범위에는 시부야강이 관통하고 있으며, 지상과 지하의 교통 결절 기능을 강화하는 지하광장을 정비하기 위해 시부야역 지구 토지구획정리사업에 의한 시부야강 이전을 실시할 필요가 있었다(pp.150-151).

또한 하천 구역에는 관민 이용을 위한 구분 지상권을 설정할 수 없기 때문에 이나리 다리의 상류부의 법령상 위치를 「하천 구역」에서 「하수도」로 변경하는 한편, 시부야 스트림 남부 철도 부지(민지)를 하천 구역으로 전환함으로써 종전 하천 부분의 부지화를 실현하고 있다(오른쪽 그림).

전환에 의해 새롭게 하천 구역으로 지정된 도요코선 선로 부지에는 강변 산책로가 정비되었다. 이나리 다리에서부터 약 600m의 산책로가 탄생되어, 시부야강 상공의 광장(pp.148-149)과 정화된 물을 활용한 「벽천(수경시설)」과 함께 도심부에 풍부한 보습과 활기를 주는 수변 공간 재생에 기여하고 있다.

왼쪽 위: 이나리 다리 광장, 가운데 위: 정비된 벽천, 오른쪽: 600m 이어진 시부야 강변 산책로, 왼쪽: 산책로에 넘치는 액티비티, 가운데 아래: 녹지의 윤택함을 느낄 수 있는 회유로

Above left: Inari Bridge Square. Above middle: Maintained wall fountain. Right: 600-m-long promenade along Shibuya River. Below left: Activities overflowing onto promenade. Below middle: Promenade full of rich greenery.

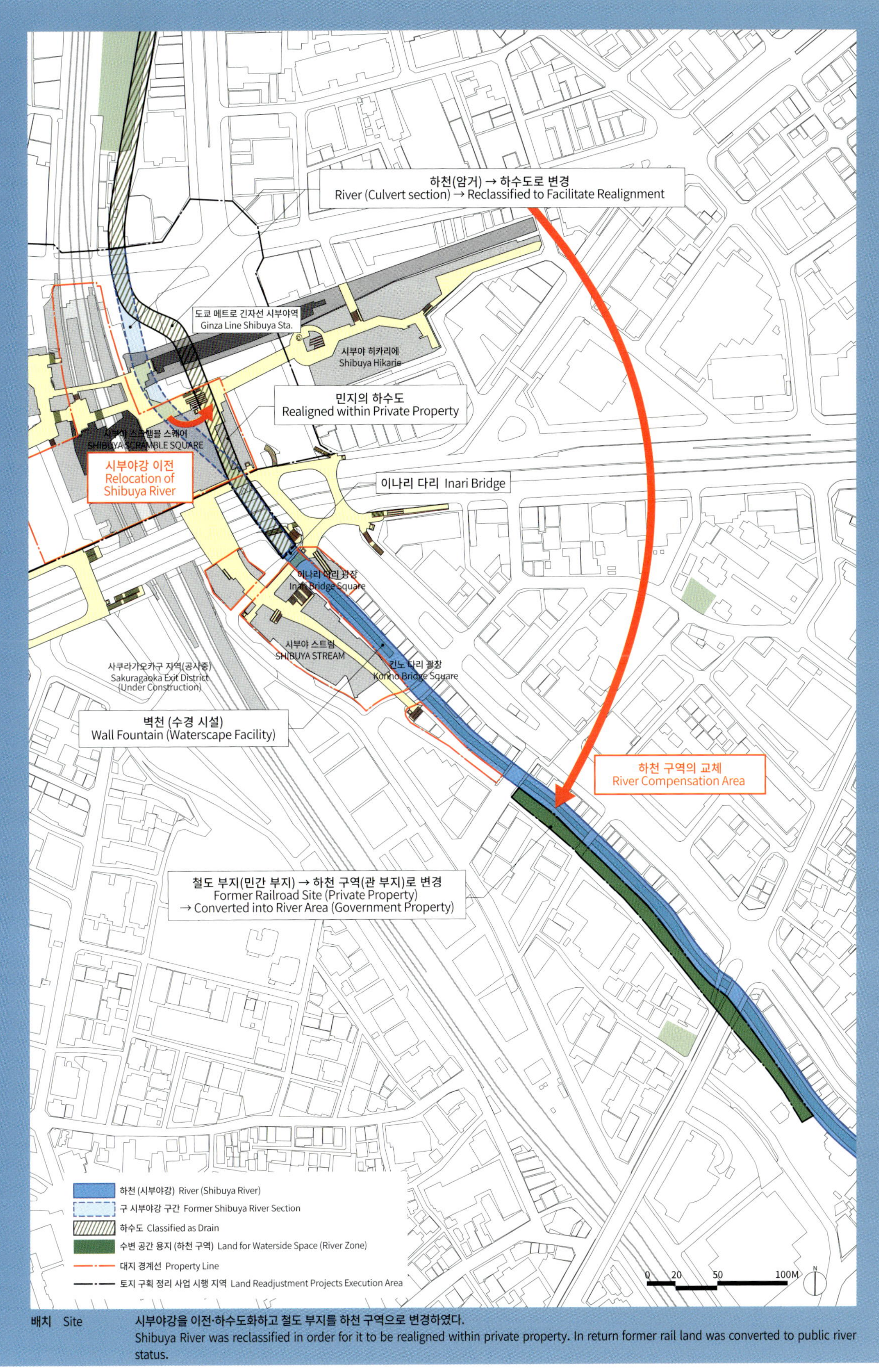

배치　Site　시부야강을 이전·하수도화하고 철도 부지를 하천 구역으로 변경하였다.
Shibuya River was reclassified in order for it to be realigned within private property. In return former rail land was converted to public river status.

Rooftop landscape connecting to the park

TOKYO MIDTOWN HIBIYA　Park View Garden

공원과 연결되는 랜드스케이프

도쿄 미드타운 히비야 파크뷰 가든(2018)

히비야 공원과 인접한 녹음이 풍부한 퍼블릭 스페이스를 저층부의 옥상에 창출하고 있다. 시설 중층에 정비된 컨퍼런스, 비지니스 지원 시설 등과 일체적인 공간을 구성함으로써 도시재생에 기여하는 효과적인 공지로 정비되었다. 도심의 유수한 녹지 공간인 히비야 공원과 고쿄 외원(外苑)을 내려다보며 쉴 수 있는 새로운 도시의 즐거움을 제공하고 있다.

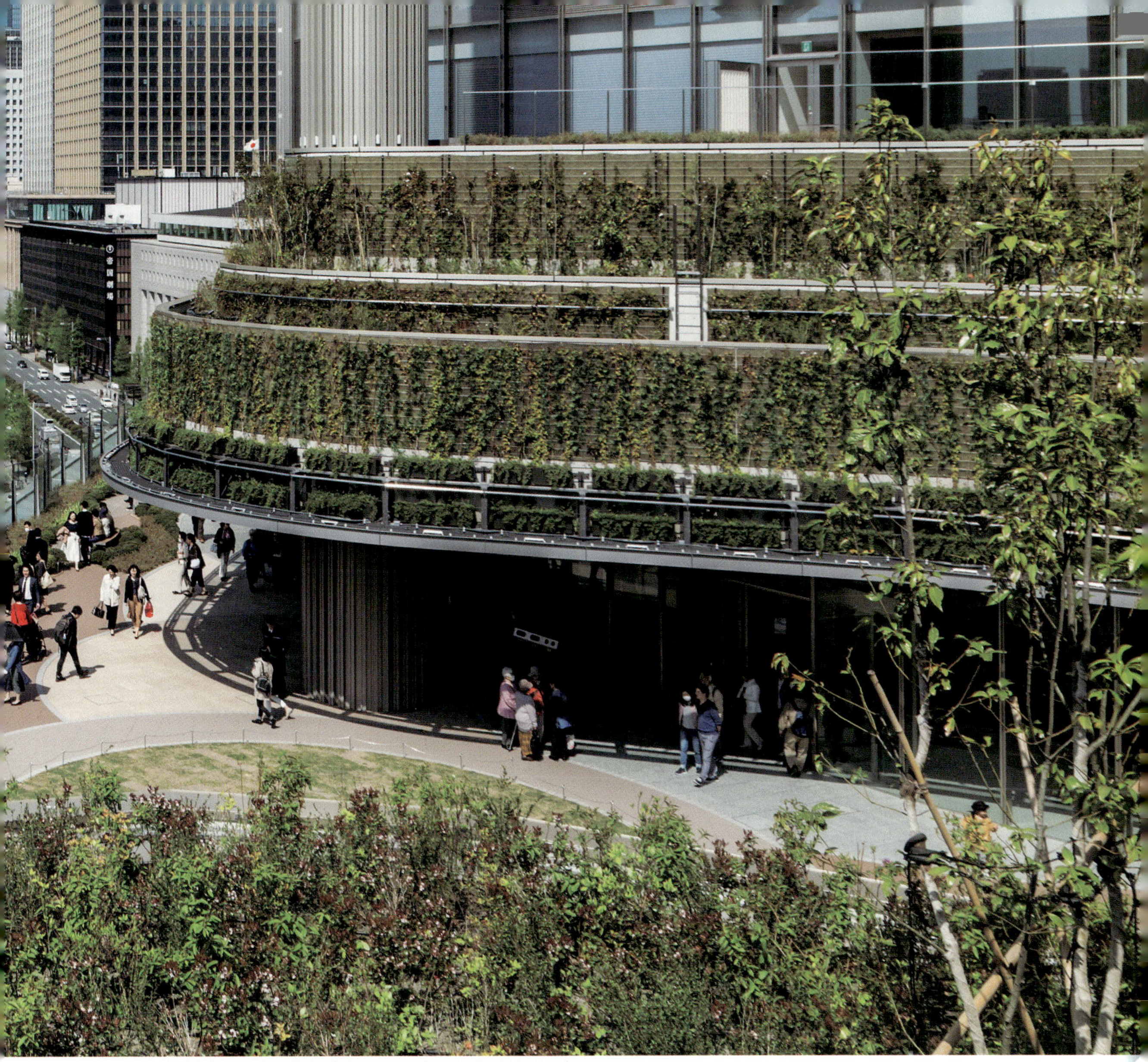

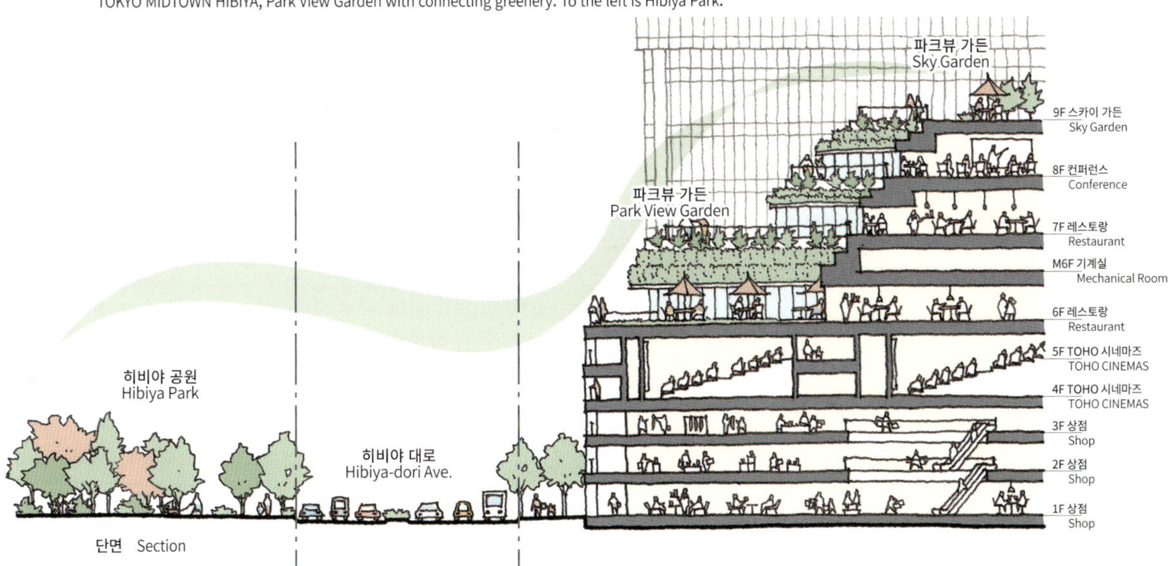

녹지가 연결되는 도쿄 미드타운 히비야 파크뷰 가든. 왼쪽으로는 히비야 공원이 보인다.
TOKYO MIDTOWN HIBIYA, Park View Garden with connecting greenery. To the left is Hibiya Park.

단면 Section

입체적으로 연결되는 광장이 다층적인 활기를 형성한다.
Vertically connected plazas create activities on multi levels.

Public space continuously connecting the building perimeter

— Shibuya PARCO · HULIC building   ROOFTOP PARK, etc.

건물 외주에 연속되는 퍼블릭 스페이스

시부야 파르코·휴릭 빌딩 옥상공원 등(2019)

업무·상업 기능을 중심으로 한 복합시설로의 재건축과 함께 건물 외주를 연속적으로 연결하여 입체적인 퍼블릭 스페이스를 창출했다. 극장 기능 등 지역의 활기를 형성하고 도시재생에 도움이 되는 공공적인 성격을 갖는 용도로 일체적인 광장이 확보되어 입체적인 활기 창출을 실현하고 있다.

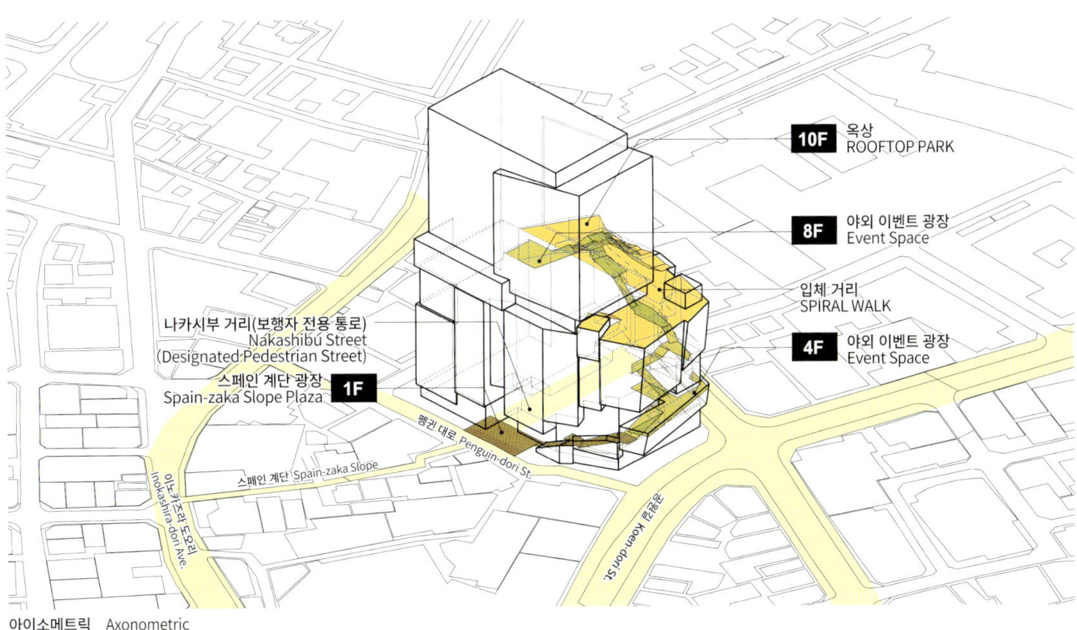

남쪽에서 바라본 관경. 10층까지 건물 외주를 나선형으로 둘러싼 입체 거리를 따라 시부야 거리의 언덕과 도로의 구성을 건물에 담았다.
View from the south. Multi-level street spirals around the building's perimeter up to 10th floor to incorporate the structures of slopes and streets of Shibuya.

아이소메트릭 Axonometric

System of PPP for renewing an urban park

— **SHIBUYAKURITSU MIYASHITA PARK**

도시 공원을 리뉴얼하는 PPP의 도입

시부야 구립 미야시타 공원(2020)

배리어 프리 동선 확보 및 시간이 지남에 따라 내진성 저하와 같은 과제를 안고 있는 시부야 구립 미야 시타 공원의 재건축에 PPP 사업을 도입하여 공원, 상업시설, 호텔, 도시계획 주차장을 일체 정비하였다. 입체 도시공원 제도의 적용을 통해 공원의 입체적인 범위를 설정함으로써 공원과 민간 시설이 동시에 정비될 수 있었다. 이로 인해 공원은 각 시설과 연계하여 다양한 액티비티가 전개되는 퍼블릭 스페이스로 거듭났다.

동쪽에서 바라본 관경. 상업시설과 공원이 적층된다.
View from the east. Commercial facilities and the park are layered.

남쪽에서 바라본 조감
Bird's eye view from the south.

입면 Elevation

시부야 구립 미야시타 공원
SHIBUYAKURITSU MIYASHITA PARK.

Mobility terminal and public space to be built above national road

National Route 15 / Shinagawa Station West Square

ANTICIPATED PROJECT

국도 상공에 만들어지는 교통 터미널과 도시광장

국도 15호 시나가와역 서쪽 출구 앞 광장

시나가와역 서쪽 출구 역 앞을 남북으로 달리는 국도 15호의 폭을 확장한 후, 그 상공에 역 앞 광장 확장과 함께 최첨단의 모빌리티 승강장을 집약한 「차세대 교통 터미널」과 사람들이 모여 쉬는 「도시광장」 등을 일체적으로 정비할 계획이 검토되고 있다. 차세대형 교통 터미널은 민관 협력에 의해, 활기찬 광장은 입체 도로 제도의 활용을 전제로 한 민간사업자에 의해 각각 정비를 실시하는 것으로 검토되고 있다.

국도 15호 시나가와역 서쪽 출구 앞 광장 이미지
Image of National Route 15 / Shinagawa Station West Square.

※ 나중에 필요한 기능 배치를 표현한 것이며, 정비 내용을 결정하는 것은 아니다.
※ 향후 관계 기관과의 협의·조정이 필요하다.

배치　Site

Revitalization of the area along Nihonbashi River by utilizing the multi-level road system

Public space along Nihonbashi River

입체 도로 제도를 활용한 니혼바시 강변 재생

니혼바시 강변 퍼블릭 스페이스

야에스 1초메 북쪽 지구 니혼바시 무로마치 1초메 지구

니혼바시에서는 노후화된 수도고속도로를 여러 개발지역과 연계해 지하화하고, 도쿄 니혼바시강을 매력적인 도시 경관으로 재생하는 프로젝트가 진행되고 있다. 입체 도로 제도를 활용하여 민간 부지 내의 수도고속도로 지하화 루트를 정비하고, 강가에는 연속적인 오픈 스페이스 창출을 유도하고 있다.

니혼바시 강변의 정비 이미지(니혼바시 무로마치 1초메 지구)
Image of the development along Nihonbashi River.

니혼바시 강변의 정비 이미지(야에스 1초메 북쪽 지구)
Image of the development along Nihonbashi River.

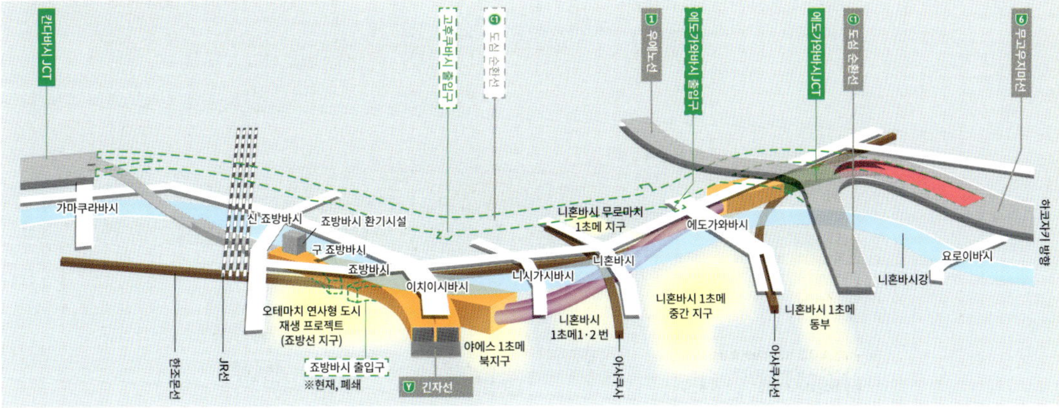

수도고속도로 지하화 루트　Undergrounded Metropolitan Expressway Route

V

AREA MANAGEMENT

**Public spaces cultivated and operated
by local communities**

**지역에서 육성하고 운영하는
퍼블릭 스페이스**

Expanding the liveliness of the street to the town
— Nihonbashi Muromachi area management

거리의 활기를 지역에 펼치다
니혼바시 무로마치 에어리어 매니지먼트(2014 ~)

니혼바시 무로마치에서는 에도 이후 계승되는 거리에 접한 상업 점포 등의 활기와 함께 건물 사이에 존재하는 도로와 광장 공간 등의 퍼블릭 스페이스를 활용해 다양한 액티비티를 만들어 니혼바시 다운 활기를 지역 전체에 펼치고 있다. 「일반사단법인 니혼바시 무로마치 지역 매니지먼트」는 니혼바시 무로마치 동부 지역 개발로 정비된 에도 벚꽃거리 지하보도(구도로)를 관리하는 중간 법인으로 발족되었다. 활동 초기에는 개발권자의 업무 위탁으로 퍼블릭 스페이스를 관리해 매번 행정 절차를 진행하고 있었기 때문에 퍼블릭 스페이스의 원활한 활용에 어려움이 있었지만, 2016년 이후 국가 전략 특구에 관한 도로법의 특례 인증을 취득해 중앙구·개발권자·본 단체 3자간 양해 각서를 체결하는 것으로, 원만하게 일체적인 퍼블릭 스페이스 활용을 실현하고 있다. 또한 현재는 개발에 의해 만들어진 민관 내의 광장 공간의 운영 위탁 업무도 담당하고 니혼바시 지역 전체에 활기 양성을 실현하고 있다.

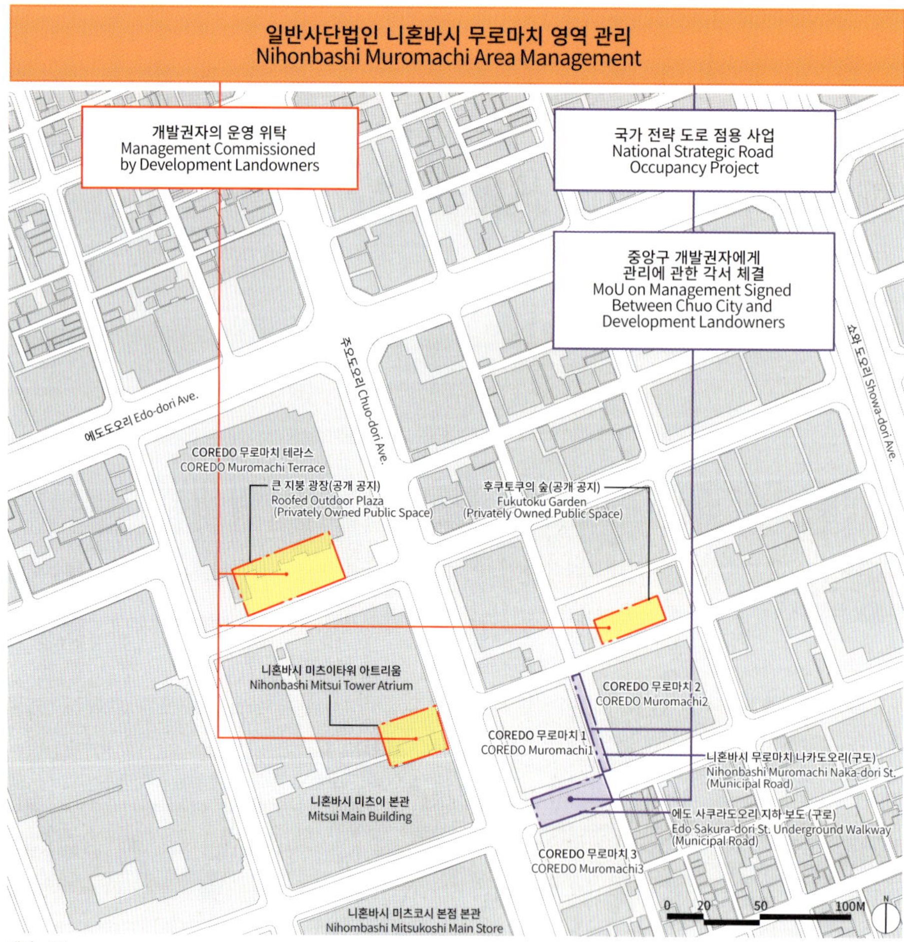

배치　Site

니혼바시 무로마치 나카도오리(구 도로 19호선의 일부)에서 이벤트가 개최된다.
Events are held on Nihonbashi Muromachi Naka-dori Street (part of Municipal Road 19).

20×60m의 유리 지붕이 있는 약 1,500㎡의 공개 공지의 활기찬 모습
Liveliness that unfolds in the approximately 1,500㎡, privately owned
public space covered by a 20×60m glass roof.

Nurture a larger community by inheriting human compassion and local atmosphere

Kanda Awajicho area management centered around WATERRAS and Awaji Park

지역의 정서와 문화를 토대로한 커뮤니티의 육성

와(和)테라스, 아와지 공원을 중심으로 한 한신 아와지 지역 매니지먼트(2012 ~)

와테라스의 저층부에는 한신 아와지 지역의 다양한 이벤트가 개최되는 퍼블릭 스페이스가 마련되어 있어 전통이 숨쉬는 지역의 교류를 육성하고 있다. 이 장소는 「아와지 공원」과 부지 내의 「광장」, 「와테라스 코몬」을 일체적으로 정비하여 만든 공간으로, 「일반사단법인 아와지 지역 매니지먼트」는 이 장소를 한신 아와지 지역의 매력 발산과 지역 커뮤니티 조성의 장

으로 활용하며 지역의 활기를 북돋우고 있다. 일반사단법인 아와지 지역 매니지먼트는 지역 활동에 학생들이 참여할 수 있는 기회를 창출하기 위해 학생 맨션 입주자가 지구 내외의 다양한 지역활동에 참여하는 구조를 구성하고 지역의 새로운 담당자 양성을 통한 지역환원 활동에 임하고 있다.

왼쪽: 와테라스의 구내 광장(앞)과 아와지 공원(안쪽)을 일체적으로 이용한 이벤트, 오른쪽 위: 와테라스 마르세 개최(광장), 오른쪽 아래: 와테라스 마르세 개최(공원)
Left: An event integrally utilizing a plaza on the WATERRAS property (front) and Awaji Park (back). Top right: Hosting of WATERRAS MARCHE (plaza). Bottom right: Hosting of WATERRAS MARCHE (park).

배치　Site

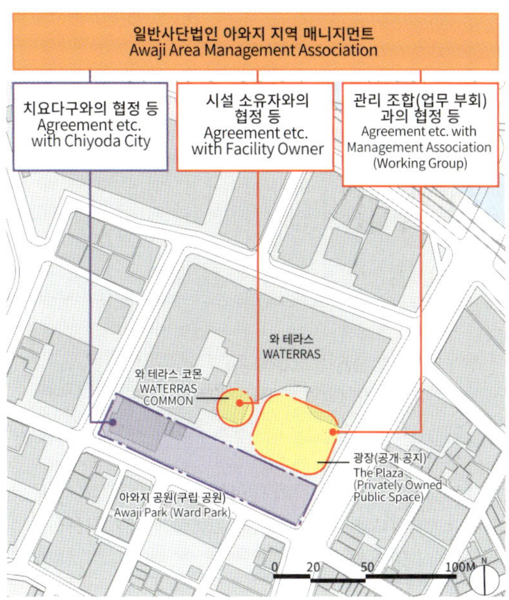

지역활동에 학생들이 참여하는 구조
Structure to drive participation of students in local activities.

Town renewal realized by utilizing public and private open space

Initiatives by Shinjuku West Area Development and Management Committee

민관 오픈 공간 활용을 통한 지역의 리뉴얼

신주쿠 부도심 지역을 대상으로 한 환경개선위원회 활동(2010~)

1970~1990년대에 개발된 초고층 빌딩이 늘어선 니시 신주쿠 지역에서는 시대의 흐름에 따른 새로운 라이프 스타일 및 워크 스타일에 대응하기 위해 다양한 활동을 연결하는 장소 만들기를 목표로 하여, 도로·공원·공개 공지를 일체적인 민관 오픈 스페이스로 설정하고 다양한 선진적인 방식을 도입하고 있다. 신주쿠 부도심 지역이라는 광대한 범위에서의 과제 해결과 도시 간 경쟁력 향상 목표를 이루기 위해, 2010년에 지역 내의 주요한 빌딩 사업자로「일반사단법인 신주쿠 부도심 구역 환경개선위원회」를 발족하고 논의를 시작했다. 그후 신주쿠구와 함께 서쪽 신주쿠 간담회를 시작하여, 향후 초고층 빌딩 저층부의 갱신과 공개 공지 활용 등을 포함하여「지역 만들기 지침」을 책정하는 등 민관 협력하에 다양한 프로젝트를 실시하여 지역 재생을 진행하고 있다.

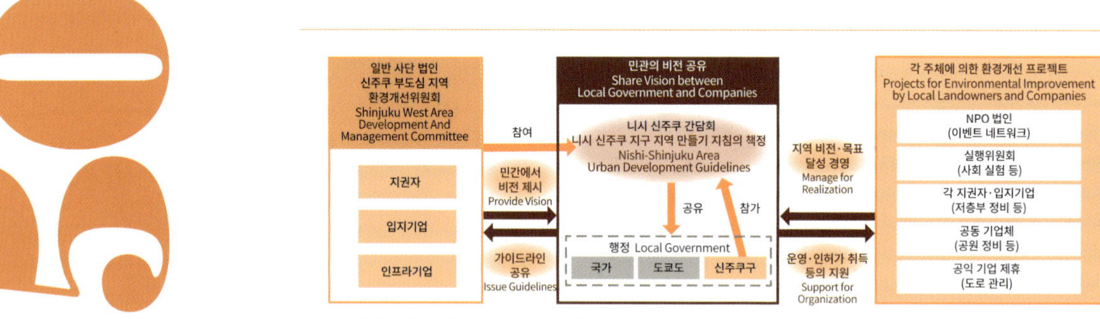

민관 협력을 위한 구조.   System for public-private partnership.

도로 위를 이용한 오픈 카페의 설치.   Open-air café was created by leveraging on-street space.

국가 전략 도로 점용 사업
National Strategic Road Occupancy Project

단면　Site

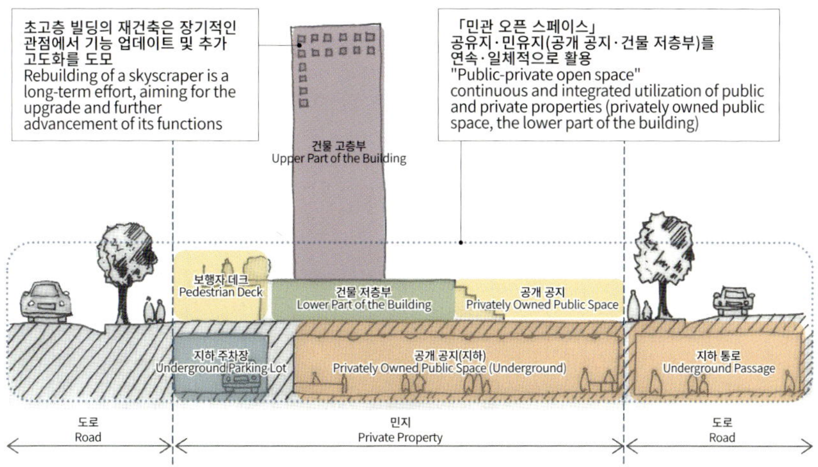

단면　Section　민관의 오픈 스페이스가 일체적으로 이용되고 있다.
Open spaces owned by public and private sectors are used in an integrated manner.

A system to operate a city precinct with a square built by the public and private sectors as its core

— Area management of Hibiya Step Square and the surrounding area

민관이 만든 광장을 중심으로 지역을 운영하는 방법

히비야 스텝 광장 및 주변 지역의 지역 매니지먼트(2018~)

영화관이나 극장이 모여 있는 히비야 지역의 중심에 위치한 이 광장은 보행자 전용 구 도로와 함께 도시에 질 높은 퍼블릭 스페이스를 제공하고 있다. 영화를 비롯한 문화·교류 도시·히비야의 역사를 아이덴티티로 하여 다양한 이벤트가 개최되고 있다. 수준 높은 정비가 이루어진 광장과 도로를 지속적으로 유지 및 관리하는 방법으로 지역 매니지먼트를 통한 관리·운영체계가 구축되었다. 개발 사업자를 포함한 현지 관계자로 조성된 「일반사단법인 일본 히비야 지역 매니지먼트」는 치요다구와의 협정에 따라 치요다구 소유지 시설의 활용에 따른 수익을 자금으로 광장 운영 및 유지관리, 구 도로 유지관리, 지역 활성화 사업 등을 실시한다.

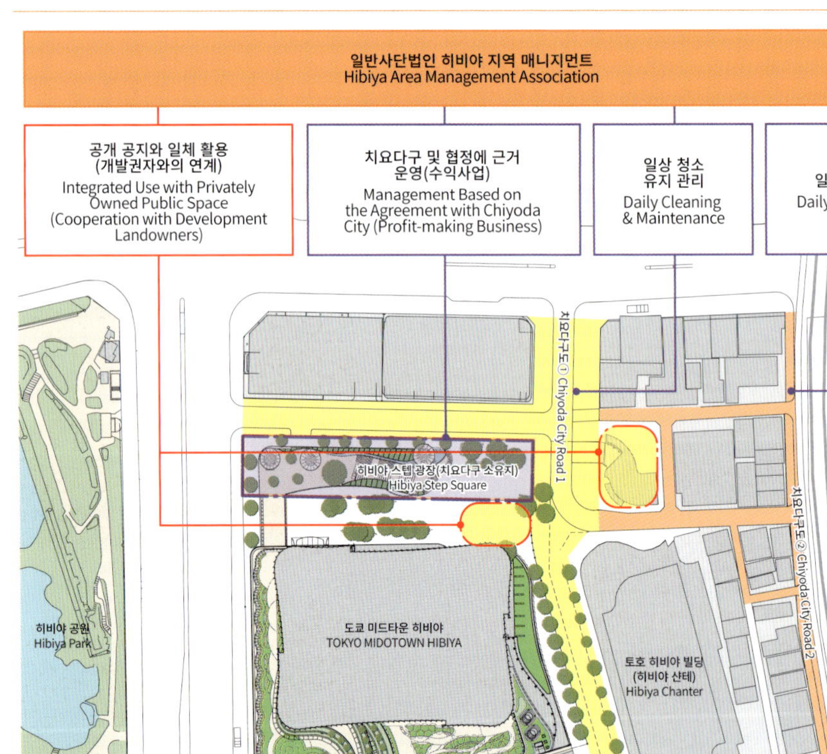

배치　Site

히비야 스텝 광장 계단에서 바라본 관경. 구 소유지와 공개 공지의 일체 활용을 도모하고 있다.
View from the steps of Hibiya Step Square. The integrated utilization of municipal property and privately owned public space is achieved.

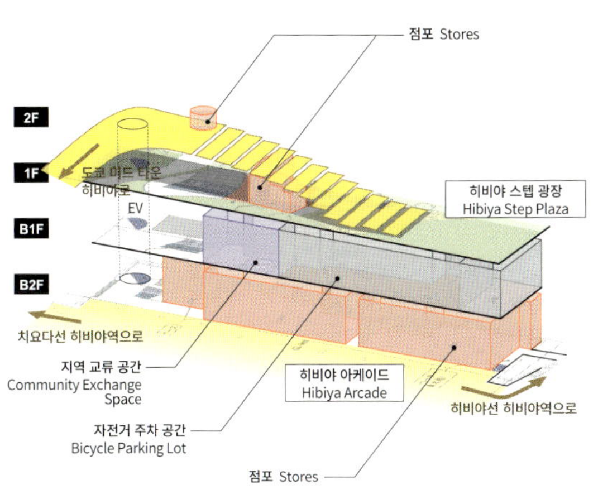

아이소메트릭　Axonometric

위: 동쪽에서 히비야 스텝 광장을 바라본 관경
아래: 히비야 스텝 광장의 계단은 이벤트 시에는 객석으로 이용된다.
Above: View of Hibiya Step Square from the east.
Below: Hibiya Step Square is used as seating during events.

마루 빌딩 내의 아트리움 「마루 큐브」. 마루노우치 나카도오리에 접한 공개 공지
Marunouchi Bldg. Atrium "MARUCUBE". Privately Owned Public Space facing Marunouchi Naka-dori Street.

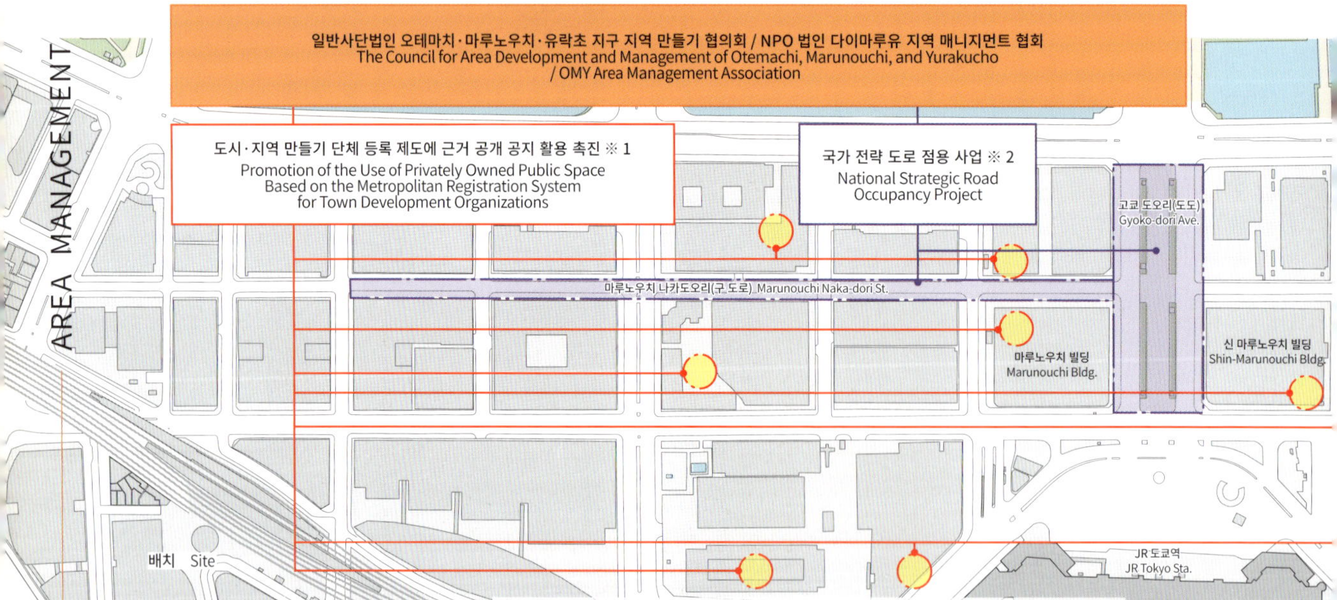

※ 1 다이마루유 지역 15 도시 구획의 공개 공지 활용을 촉진
Promoted the use of privately owned public spaces of 15 city blocks in OMY are

※2 고쿄도오리 지하 치요다 보행자 전용 도로 5호선(도도) 가와바타 녹색 도로(구 도로)도 국가 전략 도로 점용 사업 대상지로 지정
Underground of Gyoko-dori Ave. / Chiyoda Pedestrian Road No. 5 (Metropolitan Road) and Kawabata Green Corridor
(Municipal Road) are also the designated as National Strategic Road Occupancy Projects.

Bringing relaxation to the business center
—— Area management of Otemachi-Marunouchi-Yurakucho District

비즈니스 센터에 휴식을
다이마루유 지구의 지역 매니지먼트(1960년대~)

다이마루유 지구에서는 1960년대부터 민간 토지 소유주가 마루노우치 나카도오리를 비롯한 퍼블릭 스페이스의 관리 및 활성화에 힘써 왔다. 「일반사단법인 오테마치·마루노우치·유락초 지구의 지역 만들기 협의회」(1988년 설립, 2012년 사단법인화) 및 「NPO 법인 다이마루유 지역 매니지먼트 협회」(2002년 설립)는 지역의 오피스 워커나 방문객을 위한 이벤트 등을 기획·운영하고, 도로나 15개의 도시 구획에서 공개 공지에 활기를 만들고 있다. 2015년 도쿄에서는 처음으로 국가 전략 도로 점용 사업이 승인되어 차량 교통 규제를 실시하여 오픈 카페의 도입이 시작되었다. 현재는 이러한 노력이 정착되어 평일에는 사무실 워커가, 휴일에는 쇼핑객이 휴식을 취하는 모습을 볼 수 있다. 오피스 거리의 일상적인 풍경이 스트리트를 중심으로 바뀌고 있다.

마루노우치 나카도오리 활용의 변천 History of Marunouchi Naka-dori Street Utilization

1967~	• 「마루노우치 그란 마르세의 시작(1967년 ~) 나카도오리에서 푸른 하늘 도시를 개최	• Start of Marunouchi Grand Marche (1967-) Open-air market held on Naka-dori St.
	• 「런천 산책로」의 시작(1970 년 ~) 정오~오후 1시에만 교통을 규제해 보행자 천국을 실시	• Start of Luncheon Promenade (1970-) The traffic was restricted between 12 and 1 pm, so the street is only open for pedestrians
1999~	• 「동경 미레나리오」의 시작 나카 도오리를 중심으로 일루미네이션의 아치 등에 연출하여 크리스마스 시즌의 명소가 됨(2006년~ 「교토 동경」, 2012년 「동경 미치테라스」로 이름을 변경하면서 계속 실시 중)	• Start of Tokyo Millenario The area centered around Naka-dori St. was decorated with illuminated arches, making it a new attraction during the Christmas season (Undergoing name changes —Koto Tokyo since 2006 and Tokyo Michiterasu since 2012 — the event continues to this day)
2004~	• 「오픈 카페·인·마루노우치」 등의 사회 시험 개시 • 나카도오리 약 850m 구간의 보행자 공간이나 광장 테이블 의자를 설치(민지의 카페 사회 실험)	• Start of social experiments such as Open Cafe in Marunouchi • Tables and chairs were placed in the pedestrian spaces and plazas on approx. 850m Naka-dori St. (Open cafe social experiment within the private property)
2015~	• 「어반 테라스」의 시작 • 나카도오리의 일부 특정 시간대에 차량 교통 규제를 하고 차도에도 테이블·의자를 설치하여 오픈 카페로 활용 • 2015~2016년의 공적 공간 활용 시범 사업을 거쳐 2017년부터 본격 시동 시작	• Start of Urban Terrace • Restricting vehicle traffic during certain hours in a part of Naka-dori St. and placing tables and chairs on a roadway to use the space as an open café • Following a model project of public space utilization between 2015 and 2016, the full-scale operation started in 2017

나카도오리의 이벤트 「Marunouchi Street Park 2020」
Event on Naka-dori Street "Marunouchi Street Park 2020".

나카도오리의 이벤트 「마루노우치 나카도오리 어반 테라스」
Event on Naka-dori Street "Marunouchi Naka-dori Urban Terrace".

MAP p.193

53

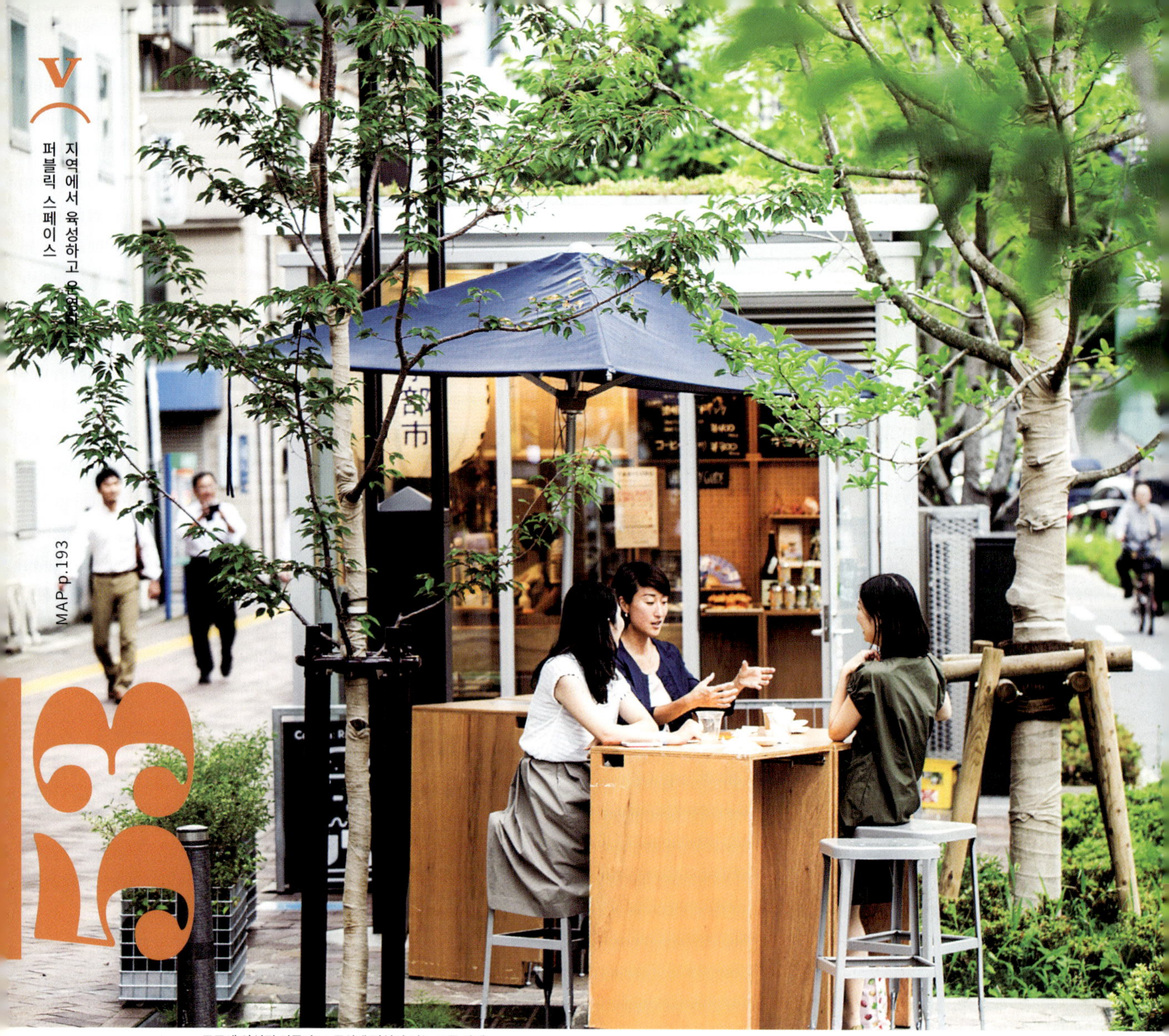

도로에 상설된 건물과 그 주위에 펼쳐져 있는 오픈 카페
Permanent building on the street and open-air café around it.

AREA MANAGEMENT

배치 Site

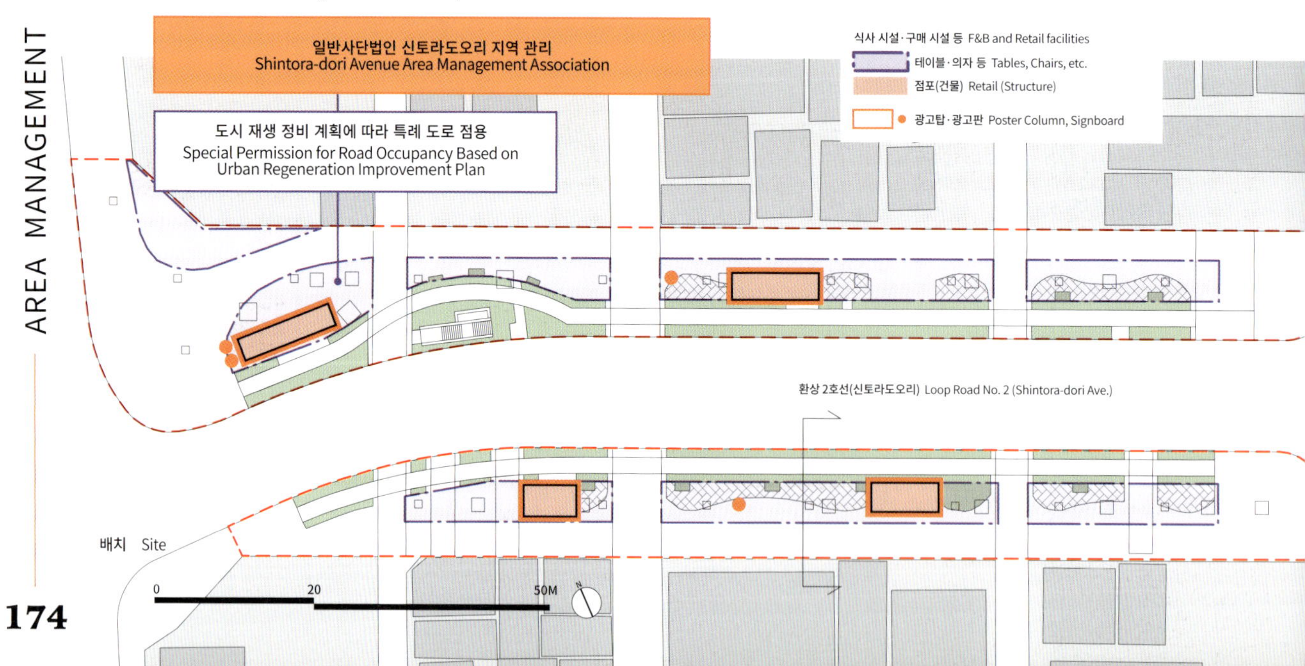

Creating liveliness with on-street architecture
—— Street utilization of Shintora-dori Avenue

도로 내 건축으로 활기 만들기
신토라도오리의 스트리트 활용 (2014 ~)

도쿄의 새로운 심볼 스트리트에서 도로 내 건축으로 활기가 생기기 시작되었다. 환상 2호선의 일부 구간으로 2014년에 개통된 신토라도오리 스트리스에서는 대로의 보도를 번화한 장소로 활용하는 「도쿄 샹젤리제 프로젝트」를 전개하여 도로 상에 오픈 카페·매장설치·이벤트 개최를 실시하고 있다. 이들은 주로 연선 권리자로 구성된 「신토라도오리의 스트리트 지역 매니지먼트 협의회」와 사업 시행자인 「일반사단법인 신토라도오리 스트리트 지역 매니지먼트」의 연계 및 협동으로 이루어지고 있다. 협의회는 거리의 경관 가이드라인을 수립하고 건물 저층부와 도로 공간으로 구성된 신토라도오리 특유의 거리 만들기를 목표로 하여 활동해 왔다. 일반사단법인은 도로 점용 허가의 특례 제도를 활용하여 도로 내 건축과 오픈 카페 설치를 가능하게 하여 거리풍경의 실현과 발전을 위해 노력하고 있다.

보도는 보행 공간의 확보도 고려하고 있다.
Ample space for pedestrians is maintained.

가로등 및 도로 내 건축의 불빛이 거리에 표정을 더한다.
Streetlights and lights of the on-street buildings give expression to the streets.

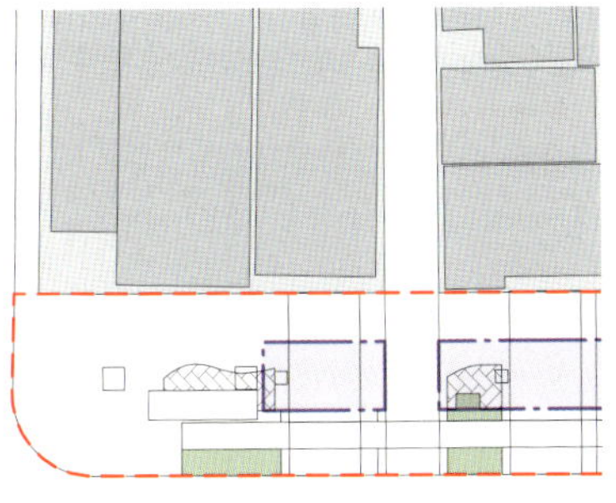

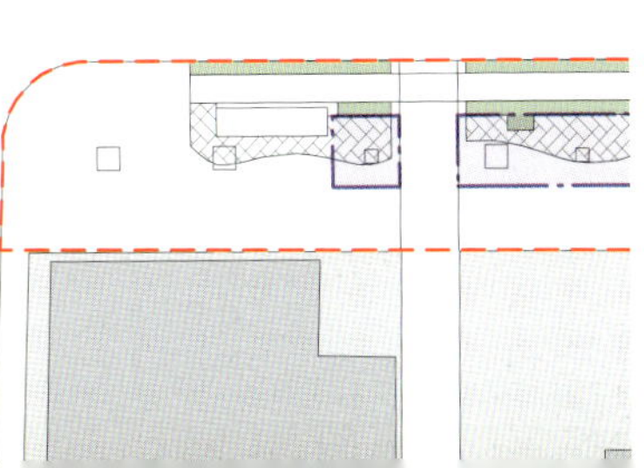

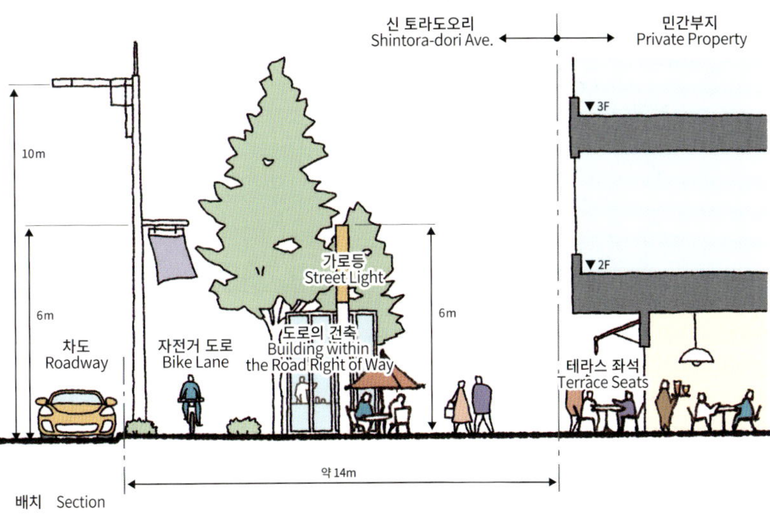

배치　Section

퍼블릭 스페이스 활용을 지원하는 제도

COLUMN

도로 공간 내에 오픈 카페 설치, 이벤트 등에 의한 활용은 도로 법률에 의거한 도로 점용 규정에 따라 일반적으로 인정받지 못했다. 2000년대에는 지역 활성화와 규제 완화의 기운이 높아져 국토 교통성 도로국에 의해 지역 활성화를 위한 노상 이벤트에 대한 알림 및 가이드라인 수립이 진행되어 각 지역에서 오픈 카페의 사회 실험이 진행되어 왔다.

이후 2011년 도시재생 특별조치법 및 도로법 시행령 개정, 2013년 국가전략 특별구역법의 제정 등을 통해 각종 지역 계획에 따라 도로에 설치하는 것이 가능하도록 제도 설계가 진행되어, 2020년 도로법 개정에서는 일정한 조건 아래에서 최장 20년의 점용도 가능해졌다. 또한 COVID-19 유행 후에는 음식점 등을 지원하기 위한 긴급조치로서 노상 음식점 등이 단체로 일괄적으로 도로를 점용하고 노상을 이용한 테이크아웃 판매나 테라스에서의 음식 제공 등을 하는 것이 인정되어 전국으로 확산되고 있다.

이러한 제도를 잘 활용하기 위해서는 도로 관리자인 행정, 교통 관리자인 경찰, 지역 활동의 주체가 되는 지역 관리 단체·반상회를 중심으로 관계자에 의한 면밀한 협의 및 조정이 필요하다. 이는 도로에 요구되는 전통적인 역할인 자동차 및 보행자 교통의 안전성과 쾌적성, 긴급 시의 피난 및 소방 등을 충족시키면서 보행자에게는 열린 공공 장소로서 활용할 수 있도록 관계자의 협동을 통해 확산되고 있다.

한편, 택지는 1960년대 이후 도시계획법 및 건축기준법에 근거하여 개발에 있어서 공개 공지의 확보를 유도하고, 도시에 부족한 퍼블릭 스페이스의 양적 확보를 추진해왔다. 도쿄도에서도 관련 법규를 「도시개발제도」라고 총칭하고, 그 운용 지침을 정해 용적률, 형태 규제 등과 함께 공개 공지 설치 기준을 정하고 있다*.

2003년에 책정된 「도쿄의 멋진 거리 만들기 추진 조례」에서 정한 「지역 만들기 단체의 등록 제도」는 공개 공지 등의 활용을 통해 지역특성을 살려 매력을 높이는 지역 만들기 활동을 주체적으로 실시하는 단체를 등록하고 그 활동을 촉진시킴으로써 민간의 발의를 끌어내면서 지역의 매력을 높이는 것을 목적으로 한 제도이다. 등록 단체는 지역의 활성화에 이바지하는 유료 이벤트를 개최하고 카페를 설치하는 등 기존보다 유연하게 개발 공터를 활용하는 것을 인정받는다. 제도 수립 이후 등록 단체는 70개를 넘어 공개 공지 정비에 대한 활용으로 움직임이 확산되고 있다.

※ 재개발 등 촉진 구역을 정하는 지구 계획, 고도 이용 지구, 특정 구획, 종합설계 4제도의 총칭. 도시재생 특별지구는 한 건당 개별 심사하기 때문에 도시개발 제반 제도에 포함되지 않지만, 공개 공지의 확보가 요구된다.

활용되는 도로 및 공개 공지의 플롯 도면
Plot diagram of utilized roads and privately owned public spaces

Nurturing the urban cultural center
— Roppongi Hills town management

문화 도심을 키우다
롯본기힐즈 타운 매니지먼트(2003 ~)

롯본기힐즈는 「문화 도심」을 테마로 문화교류 기능을 적극적으로 도입한 선구적인 대규모 개발 사례로, 롯본기힐즈 아레나는 최대 규모의 광장 공간으로 정비되었다. 주위의 건물로 둘러싸인 개폐가 가능한 대형 지붕 아래의 공간은 영화제를 비롯한 다양한 이벤트의 무대가 되어왔다. 이는 공개 공지를 건물 밖의 오픈 스페이스뿐만 아니라 활동을 유도하는 광장으로 적극 활용하고 있는 선구적인 사례이다. 롯본기힐즈의 관리 운영 주체인 모리 빌딩은 도쿄도가 정하는 공개 공지 등을 활용하는 지역 만들기 단체 등록 제도의 제 1호이다.

배치   Site

178

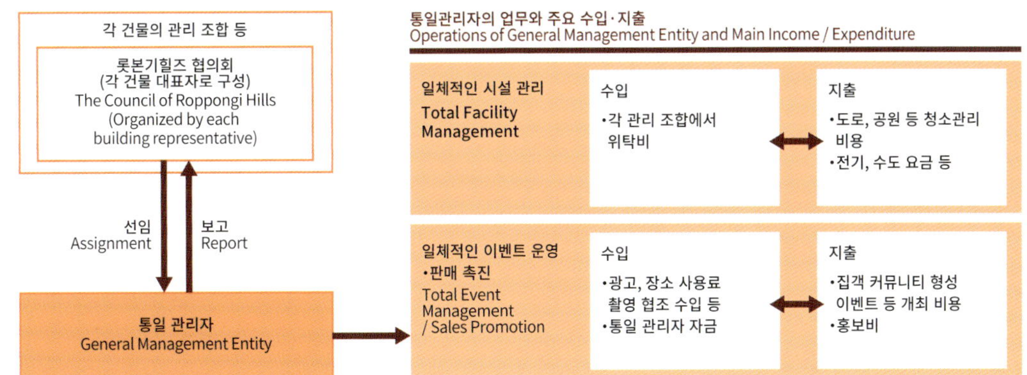

위: 롯본기힐즈 아레나는 다양한 이벤트의 무대로 이용된다.
왼쪽 아래: 지역의 앤트런스가 되는 「66 플라자」도 예술 행사에 활용된다. 오른쪽 아래: 「66 플라자」영화제
Above: Roppongi Hills Arena is used as the stage for various events. Below left: 66 Plaza, an entrance to the town, is also used for art events. Below right: Film festival.

각 건물의 관리 조합 등

롯본기힐즈 협의회
(각 건물 대표자로 구성)
The Council of Roppongi Hills
(Organized by each
building representative)

선임
Assignment

보고
Report

통일 관리자
General Management Entity

타운 매니지먼트의 구조
System of town management.

통일관리자의 업무와 주요 수입·지출
Operations of General Management Entity and Main Income / Expenditure

일체적인 시설 관리 Total Facility Management	수입	지출
	•각 관리 조합에서 위탁비	•도로, 공원 등 청소관리 비용 •전기, 수도 요금 등

일체적인 이벤트 운영 •판매 촉진 Total Event Management / Sales Promotion	수입	지출
	•광고, 장소 사용료 촬영 협조 수입 등 •통일 관리자 자금	•집객 커뮤니티 형성 이벤트 등 개최 비용 •홍보비

잔디 광장에서 요가 이벤트. 거대한 잔디 광장에서 다양한 방식으로 지역의 특징을 만든다.
Yoga event held in Grass Square. Various activities taking place on a vast lawn plaza characterizes the area.

Providing a lifestyle with vast green space

Area management of TOKYO MIDTOWN

광대한 녹색과 함께하는 라이프 스타일 제공

도쿄 미드타운 지역 매니지먼트(2007 ~)

도쿄 미드타운을 특징짓는 잔디 광장이나 미드타운 가든에서 열리는 활동은 아트 디자인 전시나 계절에 맞춘 요가, 이벤트 등 다방면에 걸쳐 있다. 고밀도 도심 속에서 느긋하게 녹색을 느끼게 하는 라이프 스타일을 제공하며, 자체 기획 홍보 이벤트뿐만 아니라 전시회, 보도 발표회 등의 수익을 얻을 수 있게 한다. 「도쿄 미드타운 매니지먼트 주식회사」는 도쿄 미트타운 지역 매니지먼트 전반을 담당하기 때문에 디벨로퍼의 출자에 의해 설립되었다. 다목적 복합개발에 있어서 전체의 통합관리를 하드 및 소프트 양면에서 실시함으로써 시설 전체의 가치 향상을 가능하게 하고 있다. 또한 2018년 이후부터 도쿄 미드타운 히비야의 일부를 운영·관리하여 업무 영역을 확대하고 있다.

상 : 도쿄 미드타운 디자인 터치 2018 PARKPACK 야외 시네마
중 : 도쿄 미드타운 디자인 터치 2018 PARKPACK
하 : 미드타운 LOVES SUMMER. 일본 여름의 더위를 테마로 한 이벤트
Above: TOKYO MIDTOWN DESIGN TOUCH 2018 PARKPACK Outdoor cinema.
Middle: TOKYO MIDTOWN DESIGN TOUCH 2018 PARKPACK. Bottom: MIDTOWN LOVES SUMMER. An event themed the cool of Japanese summer.

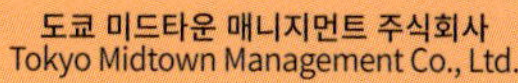

배치   Site

※ 2018년 이후에는 도쿄 미드타운 히비야의 일부 운영 · 관리를 실시한다.

운영 관리자의 주요 업무. 업무의 일환으로 공공 장소의 경영 · 관리도 하고 있다.
Main service of Operation Manager. Operation and management of the public space is included as part of the operation.

56

Fostering of creators using public space

— Utilization of privately owned public space at SHIBUYA CAST.

퍼블릭 스페이스를 활용한 크리에이터의 육성

SHIBUYA CAST.의 공개 공지 활용(2017 ~)

SHIBUYA CAST.의 콘셉트는 「삶, 일, 휴식, 다양성을 받아들이고 창의성을 유발하는 공간」이라고 되어 있다. 이 콘셉트는 퍼블릭 스페이스의 사용법과 통한다. 캣 스트리트의 입구라는 입지를 살려 건물 전면에 설치된 「GARDEN(광장·대형 계단)」과 「SPACE(다목적 공간)」을 활용하여 연간 360일에 이르는 다양한 이벤트를 개최하고 있다. 또한 시설에 거주하면서 일하는 크리에이터가 정보 발신의 기회로 공용 공간을 이용하는 「자주 기획」 활동을 지원하는 등, 퍼블릭 스페이스의 활용을 통해 다음의 시부야를 담당할 크리에이터의 육성을 도모하고 있다.

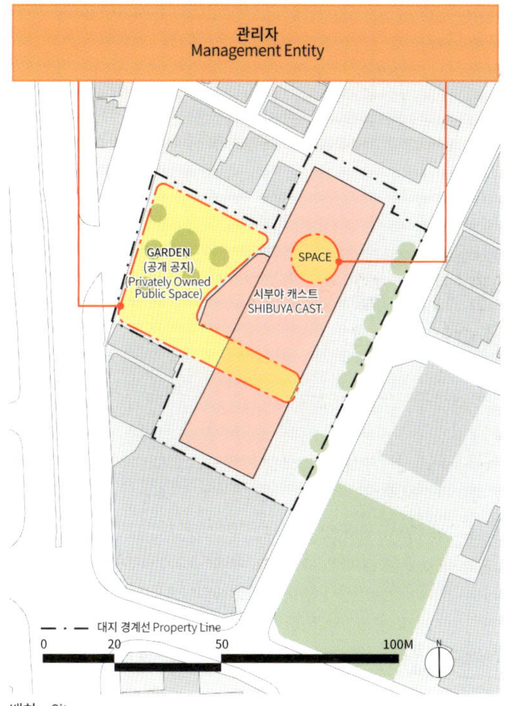

건물 저층부 서쪽에는 약 1,000㎡의 공개 공지「GARDEN」이 설치되어 다양한 이벤트가 개최되고 있다.
On the west side at the foot of the building is a large (approx 1,000m²) privately owned public garden where various events are held.

배치　Site

시부야의 새로운 즐거움을 발견하는 이벤트의 개최
An event for finding new ways to enjoy Shibuya.

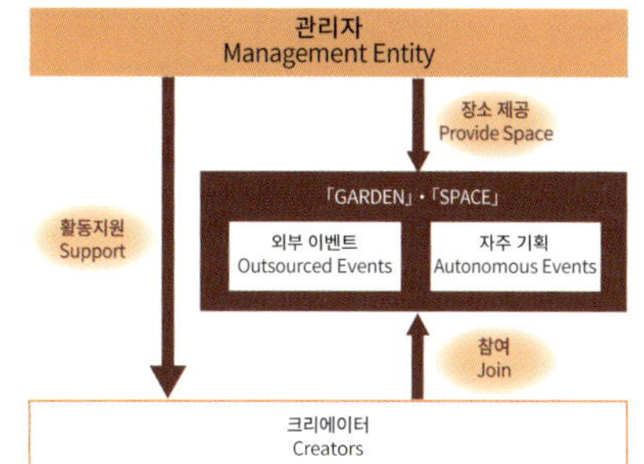

크리에이터의 활동 지원을 위한 구조
System for supporting creators' activities.

Town development unique to Shibuya leveraging outdoor advertisements

Area management in front of Shibuya Station Demonstration experiment of outdoor advertisements

옥외 광고물을 살린 시부야 다운 지역 만들기

시부야역 앞의 지역 매니지먼트 옥외 광고물 실증 실험(2015 ~)

「일반사단법인 시부야 지역 매니지먼트」가 도입한 옥외 광고물 사업에 의해 역 앞의 퍼블릭 스페이스에 새로운 정보 발신 장소가 탄생했다. 대형 광고가 전면을 향해 있고 다수의 사람이 왕래하는 도로 공간인 시부야의 스크램블 교차로는 세계적으로도 유명한 관광 명소가 되고 있으며, 이러한 「시부야 다움」을 상징하는 경관을 긍정적으로 파악해 지역 매니지먼트로 계승·발전시키는 방법이 검토되고 있다.

그 결과 새로운 제도는 다양한 기존 규칙을 완화하여 광고물 설치를 가능하게 하고, 광고로 얻은 수입을 공사 기간 동안 정보 발신 등의 지역 만들기를 위한 공익적인 자원으로 활용하게 하였다. 이로써 시부야 스크램블 광장 벽면의 대형 비전과 도로 내 광고물인 시부야 보드 시트 광고가 실현되어 시부야 다움이 있는 경관으로 지역 만들기가 진행되고 있다.

기존 규칙 Existing Rule

시부야 스크램블 광장 벽면의 대형 비전은 설치 위치·크기와 함께
기존의 규칙에서는 실현이 어렵다.
The large display on the facade of SHIBUYA SCRAMBLE SQUARE is difficult to achieve by
conventional rules in terms of both location and size

도쿄도 옥외 광고물 조례 Tokyo Metropolitan Ordinance on Outdoor Advertising Materials	도쿄도 대규모 건축물 등 경관 형성 지침 Tokyo Metropolitan Guideline for Townscape Creation Including Large-scale Buildings
• 설치 위치 : 52m 이하 • 크기 : 각 면 100m² 이내	• 설치 위치 : 3층 또는 10m 이하

새로운 규칙 New Rule

시부야 자체의 경관 형성 지침 책정
Development of Shibuya's Original Guideline for Townscape Creation

「시부야역 중심 지역 대형 건축물 등에 관련한 특정 지역 경관 형성 지침」의 변경에 의해, 시부야 다운 야간 경관이나 옥외 광고물 본연의 자세에 대해 기재되었다.

이것에 의하여 「시부야역 중심 지역 디자인 회의」에서의 협의·조정을 통해 기존의 규칙을 넘어 설치하는 것이 가능하게 되었다.

※ 일반사단법인 시부야 지역 매니지먼트 옥외 광고물 사업으로, 시부야 지역의 옥외 광고물 규칙이 마련되어 있으며 시부야 스크램블 스퀘어 비전의 광고도 그 규칙에 따른 것만 게재할 수 있다.

옥외 광고물 실증 실험의 도입
System for the outdoor advertisements demonstration experiment.

새로운 규칙에 따라 설치 가능 해졌다 시부야 스크램블 광장 벽면의 대형 비전.
The large display on the facade of SHIBUYA SCRAMBLE SQUARE, which was made possible to install thanks to the new rule.

새로운 규칙에 따라 설치 가능해졌다. 시부야 구민 헌장 보드 시트 광고(도로에 광고 물 시부야 역 하치코 광장은 행정 구역상 "도로").
The sheet advertisement (advertisement within road/Shibuya Sta. Hachiko Square is administratively designated as "road") on the Shibuya City Citizen's Charter, which was made possible to install thanks to the new rule.

Development of "theater city" in cooperation with the park
— Area management for Hareza Ikebukuro / Naka-Ikebukuro Park

공원과 연계한 「극장 도시」의 전개
Hareza 이케부쿠로·나카이케부쿠로 공원에 있어서의 지역 매니지먼트

이케부쿠로 역 주변에서는 「장르를 초월한 다양한 문화가 공존하는 세계에 유례없는 문화 예술 거점 형성」을 지역 조성 방침으로 내걸고 있으며, 역 앞 광장이나 공원의 재정비, 여러 개발 사업에서 계획 유도가 진행되고 있다. Hareza 이케부쿠로는 구 토시마구 청사 부지 등을 활용하여 사무실, 강당, 공공 시설, 공원 등으로 구성된 민관이 연계한 개발 사업으로, 본 사업의 민간 개발사업자에 의해 「일반 사단법인 Hareza 이케부쿠로 지역 매니지먼트」가 설립되었다. 이 조직은 토시마구에 의해 Hareza 이케부쿠로 지역에 위치한 나카 이케부쿠로 공원의 지정 관리자로 선정되어 유지 관리 업무를 수행함과 동시에 이벤트 공간으로서 공원의 운영을 실시하고 있으며, Hareza 이케부쿠로의 건물과 공개 공지의 대여를 일원화하는 시도를 시작하고 있다. 또한 공원 내에 있는 카페의 정비도 실시했다.

이러한 퍼블릭 스페이스와 주변 도로를 일체적으로 활용함으로써 지역 전체에 새로운 활기를 창출하고 있으며, Hareza 이케부쿠로의 콘셉트인 「모두가 주연이 되는 극장 도시」의 실현을 추진하고 있다.

배치   Site

나카이케부쿠로 공원에서 본 Hareza 이케부쿠로. 사무실, 강당, 극장, 영화관, 이벤트 공간 등으로 구성된다.
View of Hareza Ikebukuro from Naka-Ikebukuro Park. Hareza Ikebukuro consisting of a office, hall, theater, cinema, and event space, etc.

코너에 카페가 설치되는 이케부쿠로 공원. 지역 매니지먼트 조직이 지정 관리자로 선정되었다.
Cafe will be placed at a corner of Naka-Ikebukuro Park. The area management organization is selected as the designated manager.

지역으로 확대되는 퍼블릭 스페이스 / 가이드 맵

지금까지 개별적으로 소개해 온 퍼블릭 스페이스를 광역 MAP에 표시해 보면, 어느 지역들은 상호 연결되어 그 면적이 확대되는 것을 볼 수 있다. 여러 도시개발 프로젝트가 때로는 동시 다발적으로 진행되는 가운데 관계자 간의 협의와 조정을 통해 각각의 프로젝트에서 만드는 퍼블릭 스페이스가 연결되어 오랜 시간에 걸쳐 도시적인 네트워크로 발전되고 있다.

PUBLIC SPACES EXTENDING ACROSS MULTIPLE AREAS / GUIDE MAP

By collecting the projects of this issue together in maps, we see how they link up with each other and extend laterally across multiple areas.Numerous urban development projects were underway simultaneously in each area. Each project yielded new public spaces based on agreements and coordination among their various stakeholders. Over time, the spaces came to form networks at the scale of the city and, today, they support the activities of people beyond their initial intended objectives.

MAP1

도쿄역
Around Tokyo Station

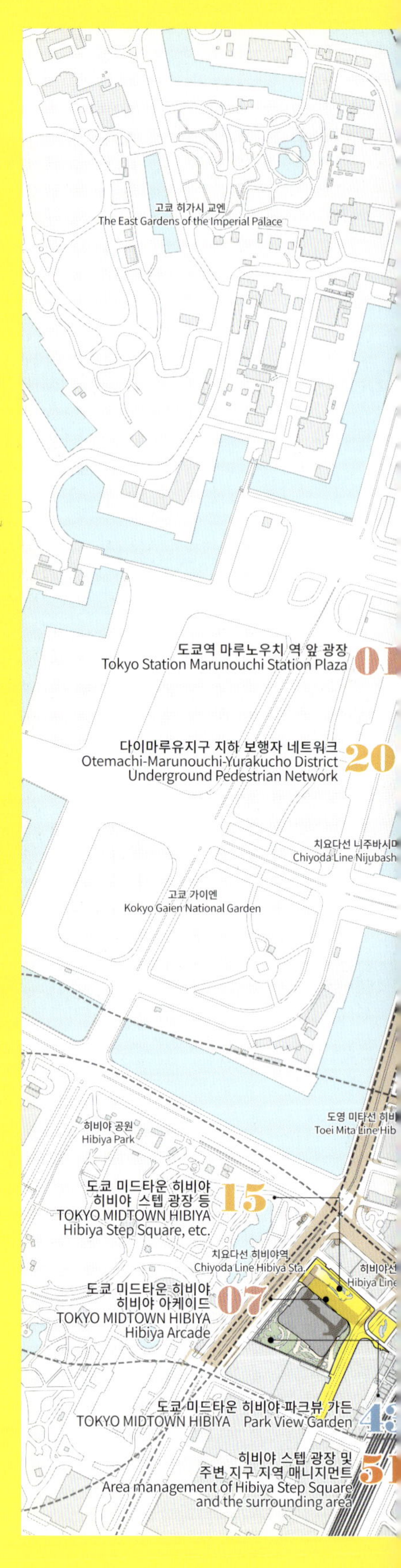

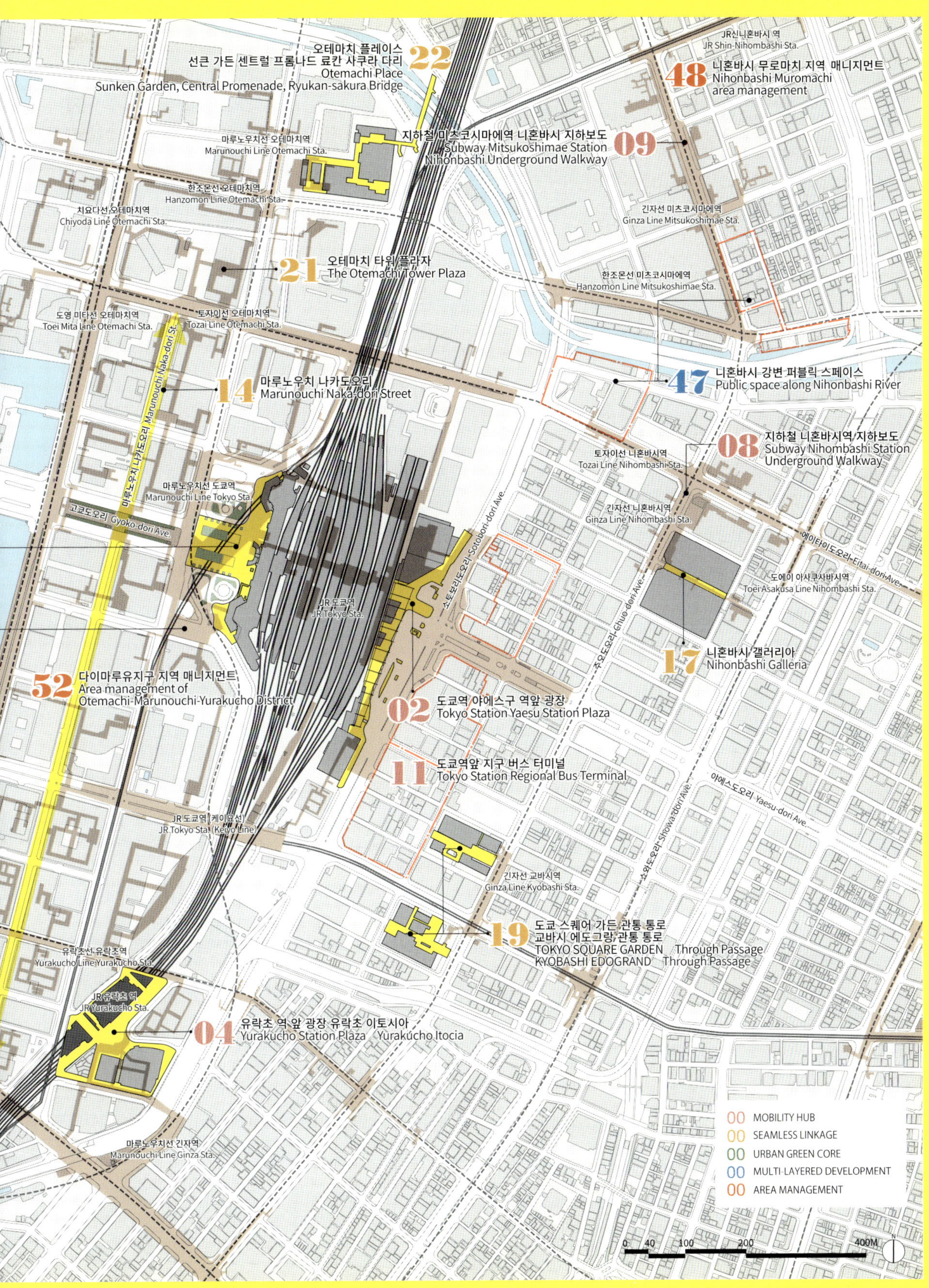

오테마치 플레이스
선큰 가든 센트럴 프롬나드 료칸 사쿠라 다리
Otemachi Place
Sunken Garden, Central Promenade, Ryukan-sakura Bridge
22

JR신니혼바시 역
JR Shin-Nihonbashi Sta.

니혼바시 무로마치 지역 매니지먼트
Nihonbashi Muromachi area management
48

마루노우치선 오테마치역
Marunouchi Line Otemachi Sta.

지하철 미츠코시마에역 니혼바시 지하보도
Subway Mitsukoshimae Station
Nihonbashi Underground Walkway

09

한조몬선 오테마치역
Hanzomon Line Otemachi Sta.

치요다선 오테마치역
Chiyoda Line Otemachi Sta.

긴자선 미츠코시마에역
Ginza Line Mitsukoshimae Sta.

21 오테마치 타워 플라자
The Otemachi Tower Plaza

한조몬선 미츠코시마에역
Hanzomon Line Mitsukoshimae Sta.

도영 미타선 오테마치역
Toei Mita Line Otemachi Sta.

토자이선 오테마치역
Tozai Line Otemachi Sta.

47 니혼바시 강변 퍼블릭 스페이스
Public space along Nihonbashi River

마루노우치 나카도오리
14 Marunouchi Naka-dori Street

토자이선 니혼바시역
Tozai Line Nihombashi Sta.

08 지하철 니혼바시역 지하보도
Subway Nihombashi Station
Underground Walkway

긴자선 니혼바시역
Ginza Line Nihombashi Sta.

에이타이도오리 Eitai-dori Ave.

마루노우치선 도쿄역
Marunouchi Line Tokyo Sta.

고코도오리 Gyoko-dori Ave.

소토보리도오리 Sotobori-dori Ave.

츄오도오리 Chuo-dori Ave.

도에이 아사쿠사선 니혼바시역
Toei Asakusa Line Nihombashi Sta.

JR 도쿄역
JR Tokyo Sta.

17 니혼바시 갤러리아
Nihonbashi Galleria

52 다이마루유지구 지역 매니지먼트
Area management of
Otemachi-Marunouchi-Yurakucho District

02 도쿄역 야에스구 역앞 광장
Tokyo Station Yaesu Station Plaza

11 도쿄역앞 지구 버스 터미널
Tokyo Station Regional Bus Terminal

쇼와도오리 Showa-dori Ave.

야에스도오리 Yaesu-dori Ave.

JR도쿄역 케이요선
JR Tokyo Sta.(Keiyo Line)

긴자선 교바시역
Ginza Line Kyobashi Sta.

19 도쿄 스퀘어 가든 관통 통로
교바시 에도그랑 관통 통로
TOKYO SQUARE GARDEN Through Passage
KYOBASHI EDOGRAND Through Passage

유락초선 유락초역
Yurakucho Line Yurakucho Sta.

JR 유락초 역
JR Yurakucho Sta.

04 유락초 역앞 광장 유락초 이토시아
Yurakucho Station Plaza Yurakucho Itocia

마루노우치선 긴자역
Marunouchi Line Ginza Sta.

00 MOBILITY HUB
00 SEAMLESS LINKAGE
00 URBAN GREEN CORE
00 MULTI-LAYERED DEVELOPMENT
00 AREA MANAGEMENT

0 40 100   200   400M

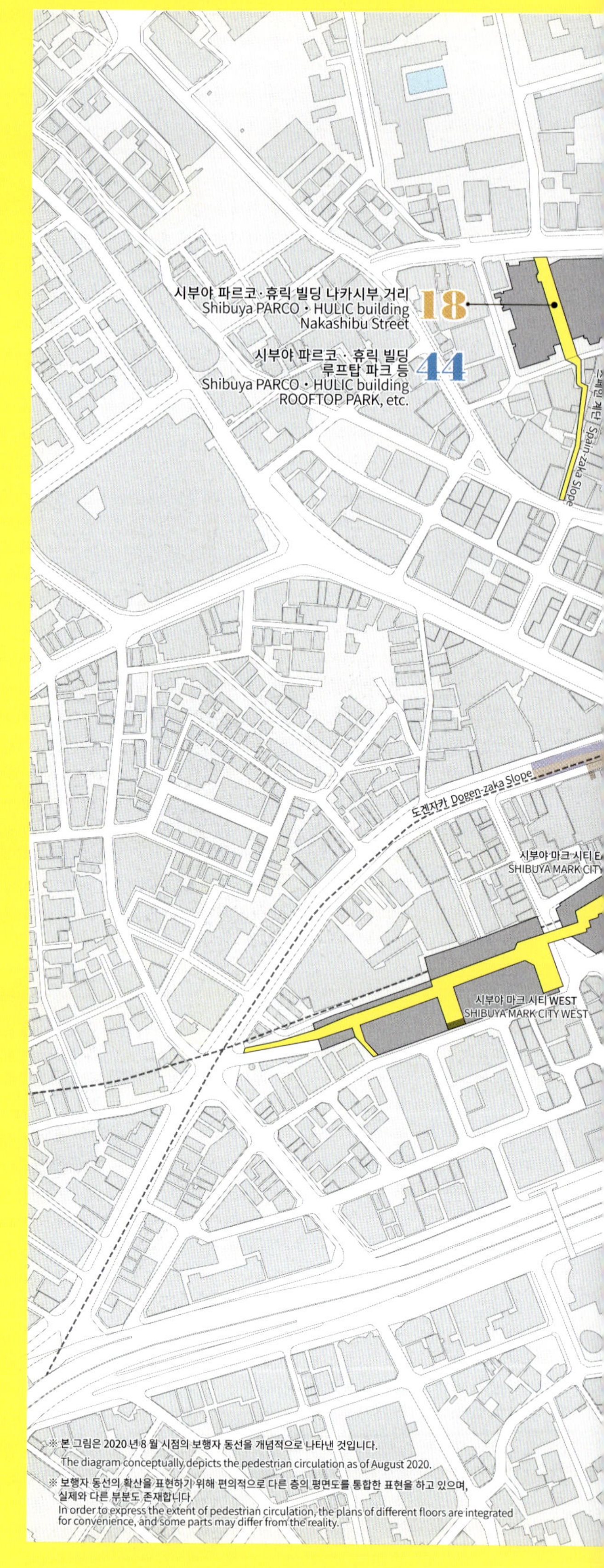

MAP2

시부야
Shibuya

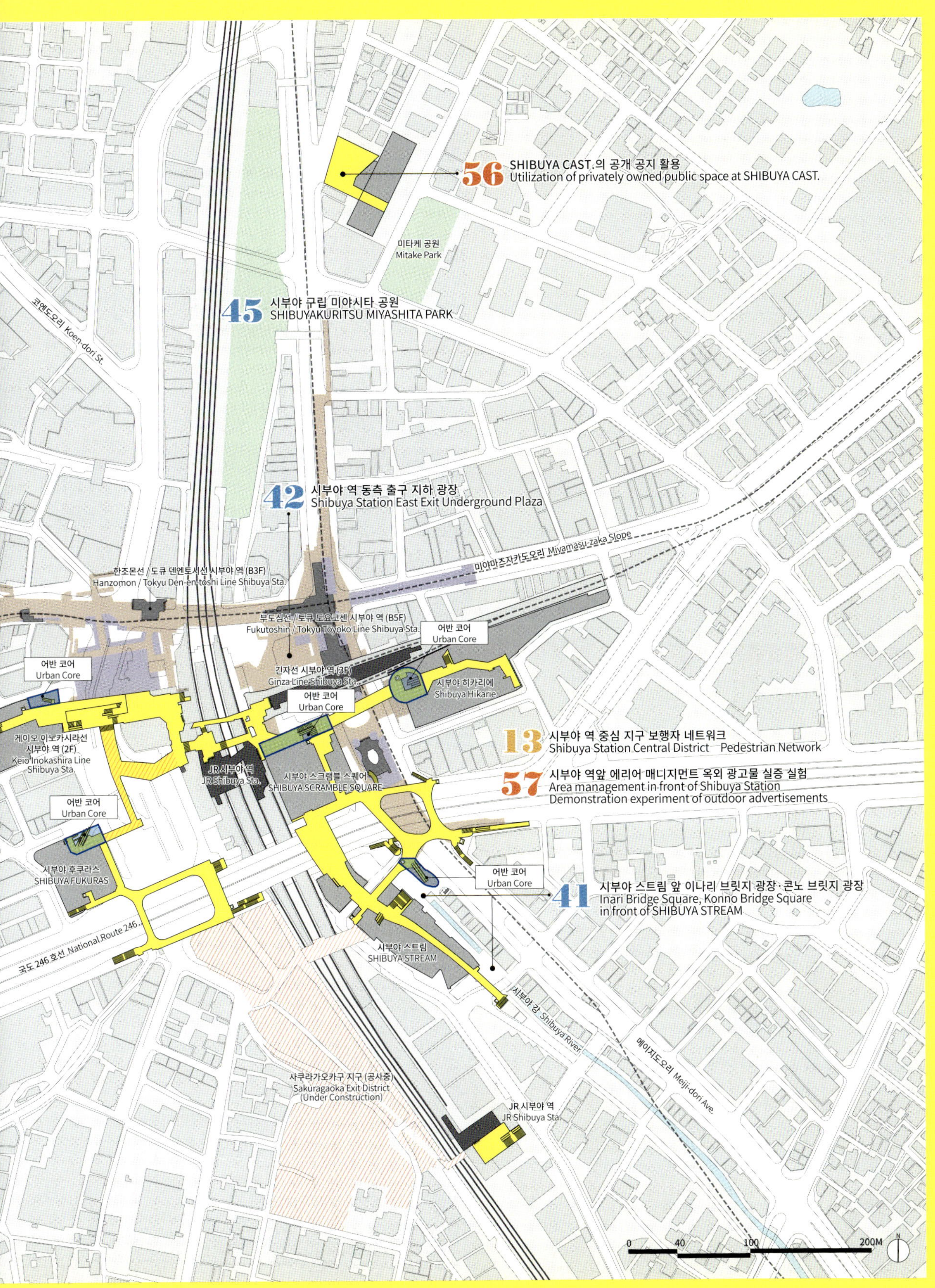

SHIBUYA CAST.의 공개 공지 활용
56
Utilization of privately owned public space at SHIBUYA CAST.
미타케 공원
Mitake Park
시부야 구립 미야시타 공원
45
SHIBUYAKURITSU MIYASHITA PARK
코엔도오리 Koen-dori St.
시부야 역 동측 출구 지하 광장
42
Shibuya Station East Exit Underground Plaza
한조몬선 / 도큐 덴엔토시선 시부야 역 (B3F)
Hanzomon / Tokyu Den-en-toshi Line Shibuya Sta.
미야마츠자카도오리 Miyamasu-zaka Slope
부도심선 / 토큐 도요코센 시부야 역 (B5F)
Fukutoshin / TokyuToyoko Line Shibuya Sta.
어반 코어
Urban Core
어반 코어
Urban Core
긴자선 시부야역 (3F)
Ginza Line Shibuya Sta.
시부야 히카리에
Shibuya Hikarie
어반 코어
Urban Core
게이오 이노카시라선
시부야 역 (2F)
Keio'Inokashira Line
Shibuya Sta.
시부야 역 중심 지구 보행자 네트워크
13
Shibuya Station Central District   Pedestrian Network
시부야 역앞 에리어 매니지먼트 옥외 광고물 실증 실험
57
Area management in front of Shibuya Station
Demonstration experiment of outdoor advertisements
JR 시부야역
JR Shibuya Sta.
시부야 스크램블 스퀘어
SHIBUYA SCRAMBLE SQUARE
어반 코어
Urban Core
어반 코어
Urban Core
시부야 스트림 앞 이나리 브릿지 광장·콘노 브릿지 광장
41
Inari Bridge Square, Konno Bridge Square
in front of SHIBUYA STREAM
시부야 후쿠라스
SHIBUYA FUKURAS
국도 246 호선 National Route 246
시부야 스트림
SHIBUYA STREAM
시부야 강 Shibuya River
에이지도오리 Meiji-don Ave.
사쿠라가오카구 지구 (공사중)
Sakuragaoka Exit District
(Under Construction)
JR 시부야 역
JR Shibuya Sta.
0   40   100   200M
N

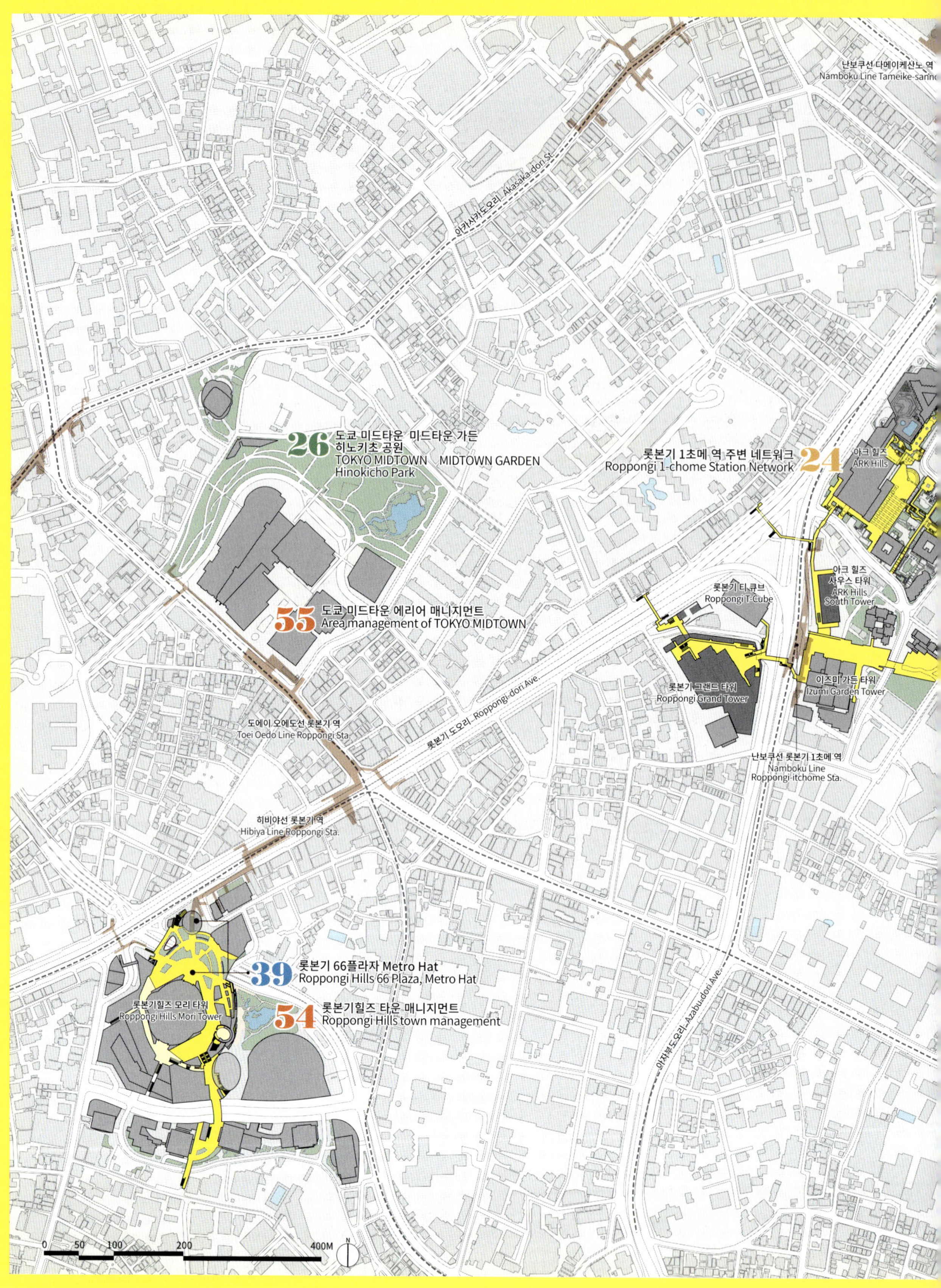
난보쿠선 다메이케산노 역
Namboku Line Tameike-sanno
도쿄 미드타운 미드타운 가든
히노키초 공원
TOKYO MIDTOWN   MIDTOWN GARDEN
Hinokicho Park
26
롯본기 1초메 역 주변 네트워크
Roppongi 1-chome Station Network
24
아크 힐즈
ARK Hills
55
도쿄 미드타운 에리어 매니지먼트
Area management of TOKYO MIDTOWN
롯본기 티큐브
Roppongi T-Cube
아크 힐즈
사우스 타워
ARK Hills
South Tower
롯본기 그랜드 타워
Roppongi Grand Tower
이즈미 가든 타워
Izumi Garden Tower
도에이 오에도선 롯본기 역
Toei Oedo Line Roppongi Sta.
롯본기 도오리 Roppongi-dori Ave.
난보쿠선 롯본기 1초메 역
Namboku Line
Roppongi-itchome Sta.
히비야선 롯본기역
Hibiya Line Roppongi Sta.
아자부도오리 Azabu-dori Ave.
39
롯본기 66플라자 Metro Hat
Roppongi Hills 66 Plaza, Metro Hat
롯본기힐즈 모리 타워
Roppongi Hills Mori Tower
54
롯본기힐즈 타운 매니지먼트
Roppongi Hills town management
아카사카도오리 Akasaka-dori St.
0 50 100   200   400M
N

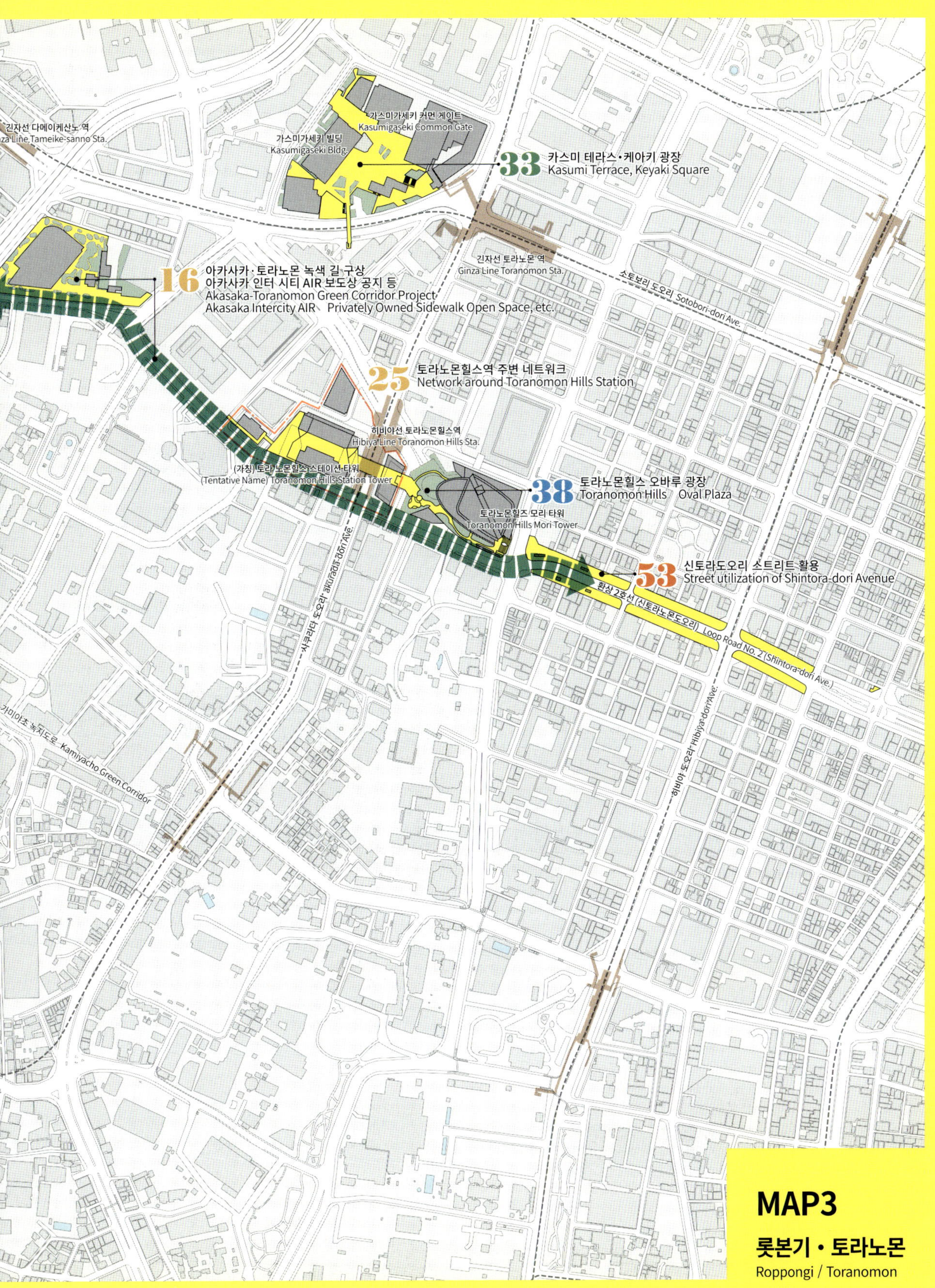

MAP3

롯본기 · 토라노몬
Roppongi / Toranomon

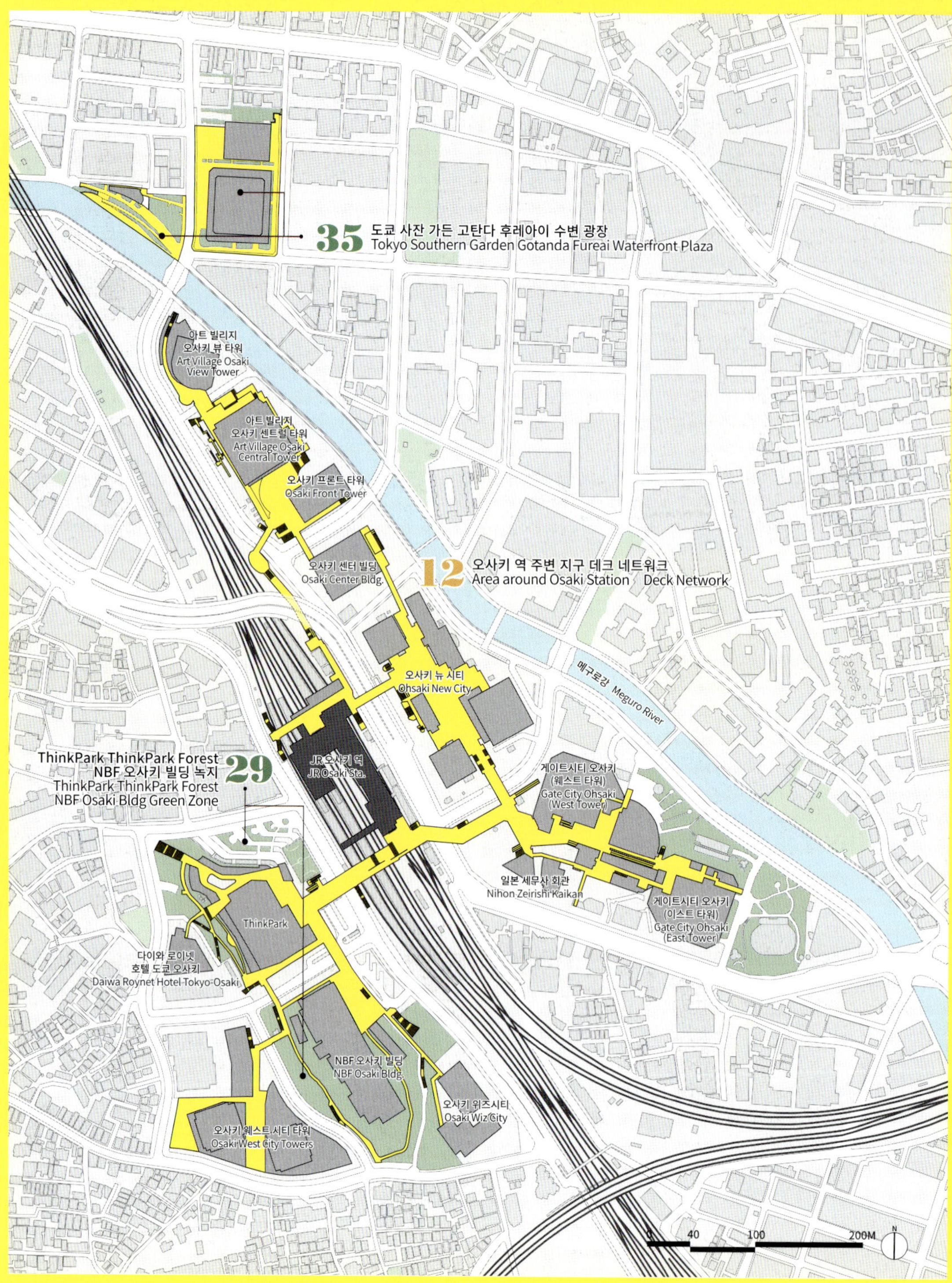

MAP4

오사키
Osaki

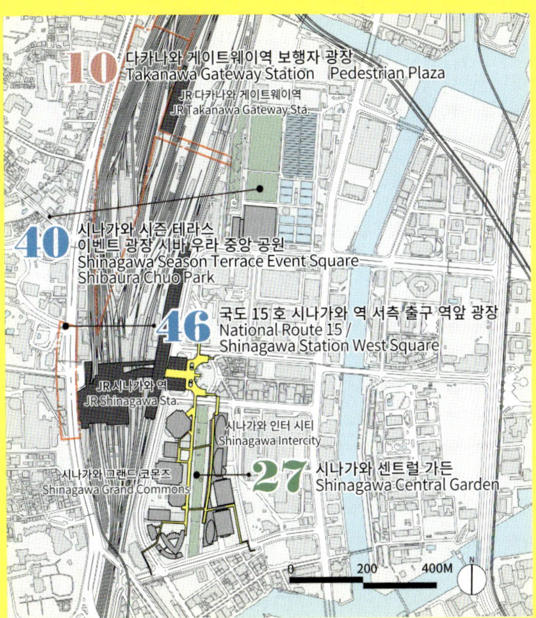

MAP5

시나가와
Shinagawa

MAP6

신주쿠
Shinjuku

MAP7

시오도메
Shiodome

MAP8

이케부쿠로
Ikebukuro

MAP9

후타고타마가와
Futakotamagawa

MAP10

오차노미즈
Ochanomizu

MAP11

오모테산도
Omotesando

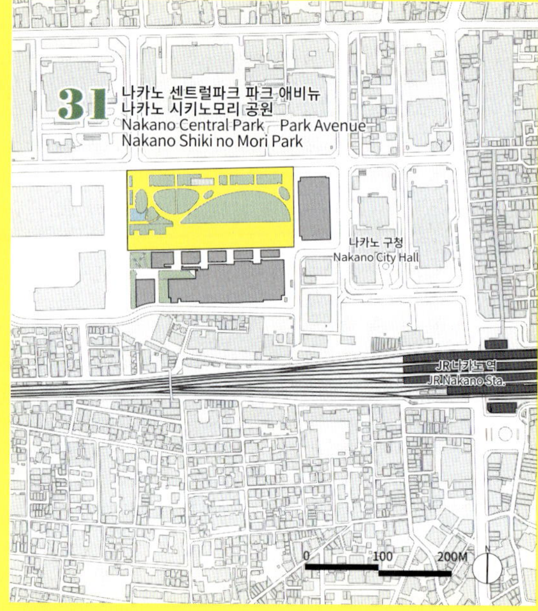

MAP12

나카노
Nakano

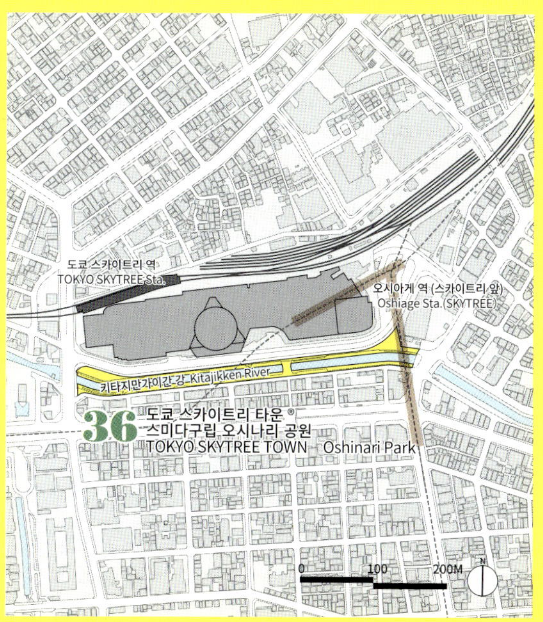

MAP13

오시아게

Oshiage

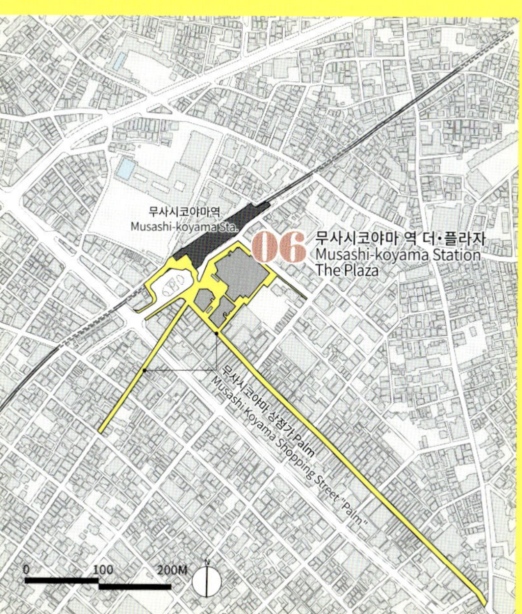

MAP14

무사시코야마

Musashikoyama

MAP15

에비스

Ebisu

MAP16

이이다바시

Iidabashi

MAP17

츠키시마

Tsukishima

【도판 출처】

- p.38 오른쪽
  국토 지리원의 항공 사진(2019)을 바탕으로 니켄세케이 작성
- pp.52-53 지하 1층 평면, p.52 정비 전후 다이어그램, p.53 지하철 미츠코시역
  앞 니혼바시 지하보도의 단면 구성
  국토교통성 도쿄 국도사무소 제공의 도면을 바탕으로 일본 설계가 작성
- p.54 CG2장
  JR 동일본 제공
- p.54 개발 후 배치
  JR 동일본 제공의 도면을 바탕으로 니켄세케이 작성
- p.55 도판 3점
  「도시 재생특별지구(야에스2초메 나카 지구) 도시계획(초안)」을 바탕으로 니
  켄세케이 작성
- p.83 개발 후 배치
  다케나카 공무점 제공의 도면을 바탕으로 니켄세케이 작성
- pp.88-89 배치
  다이마루유 지역 매니지먼트 협회 협력하에 니켄세케이 작성
- p.90 지하 2층 평면
  타이세이 건설 제공의 도면을 바탕으로 니켄세케이 작성
- p.98 완성 이미지 2점
  모리 빌딩 제공
- p.98-99 단면
  모리 빌딩 제공의 단면 CG를 바탕으로 니켄세케이 작성
- p.111 배치
  쿠메이설계 제공의 도면을 바탕으로 니켄세케이 작성
- p.112 배치
  "오사키역 서쪽 출구 지구 디자인 가이드라인"을 바탕으로 니켄세케이 작성
- p.121 배치
  KAJIMA DESIGN 제공의 도면을 바탕으로 니켄세케이 작성
- p.134 정비 이미지
  아오야마 공창
- p.135 배치
  「키타 아오야마3초메 지구(도로변 일체형 개발구역) 기본 계획」을 바탕으로
  니혼세케이 작성
- p.135 「아오야마 거리 주변 지구 마을 만들기 가이드라인」에 제시된 방향성
  「아오야마 거리 주변 지구 마을 만들기 지침」을 바탕으로 니켄세케이 작성
- p.157 아이소 메트릭
  다케나카 공무점 제공의 도면을 바탕으로 니켄세케이 작성
- p.160 도판 2장
  자료명 : 미래의 시나가와역 앞 공간(서쪽 출구) 계획~국도 15호 시나가와역
  서쪽 출구 역 앞 광장~사업 계획
  출처 : 관동 지방 정비국 홈페이지(https：//www.ktr.mlit.
  go.jp/ktr_content/content/000743189.pdf)
- p.161 강변 정비 이미지(니혼바시 무로마치 1초메 지구)
  니혼바시 무로마치 1초메 지구 시가지 재개발 준비 조합
- p.161 강변 정비 이미지(야에서 1초메 북측 지구)
  야에서 1초메 북쪽 지구 재개발 준비 조합
- p.161 수도 고속도로 지하화 루트
  수도 고속도로 주식회사의 제공 자료를 바탕으로 니혼세케이 작성
- p.169 단면
  신건축 2015년 6월 별책 「도쿄 150 프로젝트 다양한 도시 매니지먼트」 p.67
  에 게재
- pp.172-173 배치
  내각부 HP 계획도 및 다이마루유 지역 매니지먼트 협회 제공 정보를 바탕으로
  니켄세케이 작성
- p.173 마루노우치 나카도오리 활용의 변천
  「스트리트 디자인·매니지먼트 퍼블릭 스페이스를 활용하는 제도·조직·프로세
  스」(2019, 학예출판사) 등을 바탕으로 니켄세케이 작성
- pp.174-175 배치
  도시재생 정비계획을 바탕으로 니켄세케이 작성
- p.175 단면
  「신토라 도오리 경관 가이드라인(2018)」을 바탕으로 니켄세케이 작성
- p.177 도로 및 공개 공지의 플롯 도면 행정 자료를 바탕으로 니켄세케이 작성
- p.184 옥외 광고물 실증 실험의 구조
  「시부야역 중심 지역 대형 건축물 등에 관한 특정 지역 경관 형성 지침」을 바탕
  으로 니켄세케이 작성

【사진 제공】

- 신건축사
  pp.30-31, pp.32-33, pp.34-35, p.37, p.38 왼쪽, p.38 오른쪽,
  p.39, pp.42-43, p.43 아래 p.45 왼쪽, p.46 오른쪽, pp.48-49,
  p.51 2점, p.52 2점, p.62 위, p.64 2점, p.65 오른쪽, p.73 2점,
  p.77, pp.78-79, p.81, pp.82-83, p.84 아래, pp.90-91,
  p.92 2 점, p.96 2 점, pp.110-111, p113 아래 p.123 왼쪽,
  pp.124-125, p.128 2 점, p.135 상·중 오른쪽, p.142 아래
  pp.144-145, pp.148-149 후, p.148 왼쪽, pp.154-155,
  p.156, pp.156-157, p.158 2 점, p.165 2 점, p.171 중
  p.185 2 점, p.187 2 점
- photolibrary
  p.33 아래
- 니켄세케이
  p.36, p.42 아래 p.58, p.63, p.105, p.148 오른쪽, p.152 왼쪽·
  아래·위, p.181 상·중
- 일본 설계
  p.45 오른쪽, p.46 왼쪽, p.47, p.123 오른쪽, p.126, p.127,
  p.131 2 점, p.140, p.148 오른쪽, p.159, p.167 2 점
- Dick Thomas Johnson
  p.83
- 도큐
  p.45 왼쪽 위, p.183
- 시부야 스트림
  p.62 아래 p.152 왼쪽 위, 가운데 위
- 에스에스 소데 나오미치 (走出 直道)
  p.65 왼쪽
- 다이마루유 지역 매니지먼트 협회
  pp.68-69, p.70 2점, pp.172-173, p.173 2점
- 카와스미 고바야시 켄지 사진 사무실
  pp.74-75, p.106, p.107, pp.182-183
- 타이라 다케시 아틀리에
  p.84 위
- Forward Stroke inc. 포워드 스트로크
  pp.94-95, p113 위, pp.114-115, p.116 2점, p.123 위
- 도쿄 미드타운 매니지먼트
  pp.102-103, pp.180-181, p.181 아래
- 이시다 아츠시
  p.104, pp.132-133 위, p.133 오른쪽 아래
- 삿포로 부동산 개발
  pp.108-109
- 코이케 노리오
  pp.130-131 위
- 스미다구 도시 정비부 도로 공원과
  p.133 왼쪽 아래
- 아오야마 공창
  p.135에서 중간 왼쪽
- 모리 빌딩
  pp.138-139, p.141, p.142 위, p.179 위·오른쪽 아래
- 시부야역 지역 매니지먼트
  pp.150-151
- 아와지 지역 매니지먼트
  pp.166-167, p167 아래
- 신주쿠 부도심 지역 환경 개선위원회
  p.168
- 미쓰이 부동산
  p.171 위, 아래
- 신토라도오리 지역 매니지먼트
  pp.174-175, p.175 2 점
- 롯본기 아트 나이트 실행위원회
  p.179 왼쪽 아래

※특기없는 삽화는 니켄세케이, 니혼세케이에서 작성
※ 본지에 게재된 도면·일러스트는 편의상 각 시설의 이미지를 통합한 표현을 하고 실제와 다른
부분도 존재한다.

【작성자】

공익재단법인 도시 만들기 퍼블릭 디자인 센터
퍼블릭 스페이스 연구회

———

퍼블릭 스페이스 연구회
【편집 위원】
키시이 타카유키　　　계량계획연구소 대표이사
타카미 키미오　　　　호세이 대학 교수
데쿠치 아츠시　　　　도쿄대학 대학원 교수
나카이 히로시　　　　도쿄대학 대학원 교수

【사무국·집필담당】
오오마츠 아츠시　　　니켄세케이 사장
타지마 야스시　　　　니혼세케이 이사 도시계획 담당
키노시타 미즈노　　　전·도시 만들기 퍼블릭 디자인 센터
타케우치 칸사　　　　니켄세케이 고문
이토 마사토　　　　　니켄세케이 퍼블릭 에셋 라보
다나카 야스노리　　　니혼세케이 도시 계획 군
이시구로 마사유키　　니켄세케이 재개발계획부
우라타 히로히코　　　니켄세케이 퍼블릭 에셋 라보
오카모토 나루미　　　니켄세케이 퍼블릭 에셋 라보
가네마루 쇼　　　　　니켄세케이 도시계획부
강 인내　　　　　　　니켄세케이 퍼블릭 에셋 라보
사토 아츠시　　　　　니켄세케이 퍼블릭 에셋 라보
타케야마 나미　　　　니켄세케이 퍼블릭 에셋 라보
나카지마 나오야　　　니켄세케이 계획부
나카니시 히로　　　　니켄세케이 퍼블릭 에셋 라보
히라 누이　　　　　　니혼세케이 홍보실
후지와라 켄야　　　　니켄세케이 퍼블릭 에셋 라보
혼마 유리　　　　　　니혼세케이 도시계획 군
미츠야마 아카네　　　니켄세케이 도시개발부
요코세 토모히코　　　니켄세케이 퍼블릭 에셋 라보

【집필 협력】
콘노 히로키　　　　　니켄세케이 도시개발부
이와이 토무　　　　　니켄세케이 공공영역디자인부
이와나가 케이죠　　　니혼세케이 도시계획 군
우에스기 아키히토　　니켄세케이 퍼블릭 에셋 라보
우치야마 다카시　　　니켄세케이 도시개발부
왕 쇼　　　　　　　　니켄세케이 도시디자인부
오오이시 마유카　　　니켄세케이 랜드스케이프부
카라사와 유키토　　　니켄세케이 퍼블릭 에셋 라보
카케야마 료　　　　　니켄세케이 도시개발부
킨 치카　　　　　　　니켄세케이 계획부
쿠보 토모다케　　　　니혼세케이 에리어 디자인부
쿠라이시 유타　　　　니켄세케이 재개발계획부
쿠루마도 고우스케　　니켄세케이 퍼블릭 에셋 라보
황 준기　　　　　　　니켄세케이 계획부
시노즈카 고우이치로　니켄세케이 퍼블릭 에셋 라보
사사키 쇼시　　　　　니혼세케이 도시계획 군
사노 히로시　　　　　니혼세케이 도시계획 군
시부야 켄타　　　　　니켄세케이 도시계획부
시마다 카토　　　　　니혼세케이 도시계획 군
쇼 이쿠에　　　　　　니켄세케이 도시개발부
세가와 아스나　　　　니켄세케이 기획개발부
소노 요시미　　　　　니켄세케이 재개발계획부
다카하시 마리　　　　니켄세케이 재개발계획부
다카하시 요우스케　　니혼세케이 도시계획 군
타지 야스다카　　　　니켄세케이 도시개발부
타츠케 료　　　　　　니혼세케이 도시계획 군
타치이시 신야　　　　니혼세케이 도시계획 군
다나카 켄스케　　　　니혼세케이 도시계획 군
테시마 오사무　　　　니혼세케이 도시계획 군
나카무라 게이고　　　니혼세케이 도시계획 군
니시 다이스케　　　　니켄세케이 랜드스케이프부
하가 켄지　　　　　　니혼세케이 도시 계획 군
하시모토 쇼이치로　　니켄세케이 도시디자인 부
하라 이즈미　　　　　니켄세케이 도시디자인 부
마에다 나오야　　　　니켄세케이 도시개발부

마츠무라 다쿠미　　　니혼세케이 도시계획 군
미야모토 유타　　　　니켄세케이 재개발계획부
야마자키 노부히사　　니혼세케이 랜드스케이프·도시 기반 디자인부
야마다 치나　　　　　니켄세케이 종합 연구소 신 영역 부문

【일러스트 담당】
야마다 마사아키　　　니켄세케이 일러스트레이션 스튜디오
나카무라 유코　　　　니켄세케이 일러스트레이션 스튜디오
이시이 쇼이치로　　　니켄세케이 일러스트레이션 스튜디오
카라이지에후 시몬　　니켄세케이 일러스트레이션 스튜디오
가스테루 톰　　　　　니켄세케이 일러스트레이션 스튜디오
키리하라 다케시　　　니켄세케이 일러스트레이션 스튜디오
사이타 유카고　　　　니켄세케이 일러스트레이션 스튜디오
나카무라 하즈키　　　니켄세케이 일러스트레이션 스튜디오
후지나와 사나　　　　니켄세케이 일러스트레이션 스튜디오
야마미 산타　　　　　니켄세케이 일러스트레이션 스튜디오

【도판 작성 담당】
니시우 미나코　　　　니켄세케이 도시개발그룹
니혼세케이 홍보실

———

【영문교정 담당】
데미안 페이션스 니켄세케이 개발PM부

58 퍼블릭 스페이스 in 도쿄
58 Public Spaces in Tokyo

초판 1쇄 인쇄 2022년 03월 25일
초판 1쇄 발행 2022년 03월 30일

—

편저자 도시 만들기·퍼블릭 디자인 센터 퍼블릭 스페이스 연구회
역자 정병균 김미화
펴낸이 김호석
기획편집 곽유찬 권순현 박선영
디자인 전영진
마케팅 오중환
경영관리 박미경
영업관리 김경혜

—

펴낸곳 도서출판 대가
주소 경기도 고양시 일산동구 장항동 776-1 로데오메탈릭타워 405호
전화 02) 305-0210
팩스 031) 905-0221
전자우편 dga1023@hanmail.net
홈페이지 www.bookdaega.com

—

ISBN 978-89-6285-354-4 (93530)

• 파손 및 잘못 만들어진 책은 교환해드립니다.